CHAPMAN & HALL/CRC APPLIED MATHEMATICS
AND NONLINEAR SCIENCE SERIES

Introduction to
non-Kerr Law Optical Solitons

Published Titles

Computing with hp-ADAPTIVE FINITE ELEMENTS: Volume I One and Two Dimensional Elliptic and Maxwell Problems, Leszek Demkowicz

CRC Standard Curves and Surfaces with Mathematica®*: Second Edition,* David H. von Seggern

Exact Solutions and Invariant Subspaces of Nonlinear Partial Differential Equations in Mechanics and Physics, Victor A. Galaktionov and Sergey R. Svirshchevskii

Geometric Sturmian Theory of Nonlinear Parabolic Equations and Applications, Victor A. Galaktionov

Introduction to Fuzzy Systems, Guanrong Chen and Trung Tat Pham

Introduction to non-Kerr Law Optical Solitons, Anjan Biswas and Swapan Konar

Introduction to Partial Differential Equations with MATLAB®, Matthew P. Coleman

Mathematical Methods in Physics and Engineering with Mathematica, Ferdinand F. Cap

Optimal Estimation of Dynamic Systems, John L. Crassidis and John L. Junkins

Quantum Computing Devices: Principles, Designs, and Analysis, Goong Chen, David A. Church, Berthold-Georg Englert, Carsten Henkel, Bernd Rohwedder, Marlan O. Scully, and M. Suhail Zubairy

Forthcoming Titles

Computing with hp-ADAPTIVE FINITE ELEMENTS: Volume II Frontiers: Three Dimensional Elliptic and Maxwell Problems with Applications, Leszek Demkowicz, Jason Kurtz, David Pardo, Maciej Paszynski, Waldemar Rachowicz, and Adam Zdunek

Mathematical Theory of Quantum Computation, Goong Chen and Zijian Diao

Mixed Boundary Value Problems, Dean G. Duffy

Multi-Resolution Methods for Modeling and Control of Dynamical Systems, John L. Junkins and Puneet Singla

Stochastic Partial Differential Equations, Pao-Liu Chow

CHAPMAN & HALL/CRC APPLIED MATHEMATICS
AND NONLINEAR SCIENCE SERIES

Introduction to non-Kerr Law Optical Solitons

Anjan Biswas

Delaware State University
Dover, DE, U.S.A.

Swapan Konar

Birla Institute of Technology
Mesra Ranchi, India

CRC Press
Taylor & Francis Group
Boca Raton London New York

CRC Press is an imprint of the
Taylor & Francis Group, an **informa** business

A CHAPMAN & HALL BOOK

First published 2007 by Chapman & Hall

Published 2019 by CRC Press
Taylor & Francis Group
6000 Broken Sound Parkway NW, Suite 300
Boca Raton, FL 33487-2742

© 2007 by Taylor & Francis Group, LLC
CRC Press is an imprint of Taylor & Francis Group, an Informa business

First issued in paperback 2019

No claim to original U.S. Government works

ISBN-13: 978-0-367-45336-7 (pbk)
ISBN-13: 978-1-58488-638-9 (hbk)

Library of Congress Cataloging-in-Publication Data

Biswas, Anjan, Dr.
    Introduction to non-Kerr law optical solitons / by Anjan Biswas and Swapan Konar.
        p. cm. -- (Chapman & Hall/CRC applied mathematics and nonlinear
        science series)
    Includes bibliographical references and index.
    ISBN 1-58488-638-2 (alk. paper)
    1. Solitons. 2. Nonlinear waves. 3. Nonlinear optics. 4. Optical communications. I. Konar, Swapan. II. Title. III. Series.

QC174.26.W28B56 2006
530.12'4--dc22
        2006049558

Visit the Taylor & Francis Web site at
http://www.taylorandfrancis.com

and the CRC Press Web site at
http://www.crcpress.com

Preface

Recent years have witnessed an explosion of research activities in the field of soliton propagation in nonlinear optical media. These activities are motivated by the fact that optical solitons, both temporal and spatial variety, do have practical relevance in the latest communication technology based on generation and transportation of localized optical pulses. To date, these optical pulses have already made tera bit/s transmission through optical fibers feasible in the laboratory. In addition, they also play significant roles in several other technologically relevant aspects, such as optical switching and signal processing.

Early investigations in these areas began with Kerr law nonlinearity in which the refractive index of the medium is proportional to the light intensity. With further development of the subject, several forms of nonlinearity have come under investigation. Notable among these forms of nonlinearity are parabolic or cubic-quintic, power law, dual-power law, and saturating nonlinearities. These nonlinearities reveal many new and interesting behaviors hitherto unknown in Kerr law of nonlinearity. However, in spite of tremendous progress in these areas in the last 2 decades, and despite the fact that several important and extremely well-written books are now available that deal with soliton propagation in optical communication systems and allied areas, no textbook of worth deals exclusively with soliton propagation in media that possess non-Kerr law nonlinearities. Thus, to bridge the gap between availability and nonavailability, we felt the need to bring out a book exclusively dealing with optical soliton propagation in non-Kerr law media.

This book is organized as follows: Chapter 1 presents an introduction to the field of fiber optics and basic features of fiber-optic communications. The nonlinear Schrödinger's equation (NLSE) has been introduced and mathematical aspects, including conserved quantities, have been outlined in chapter 2. In this chapter, we have also introduced the perturbation to the NLSE. Adiabatic dynamics of soliton parameters have also been introduced. Finally, we have discussed the concept of quasi-stationary solitons and their influences in this chapter. In chapter 3, we have derived the NLSE for Kerr law nonlinearity from basic principle. The inverse scattering transform has been outlined and, using this principle, the 1-soliton solution has been obtained in this chapter. In addition, we have explained the variational principle and Lie transform, which are used to integrate the NLSE with Hamiltonian-type perturbation. The non-Kerr law solitons have been discussed in chapters 4 through 7. Chapters 4, 5, 6, and 7 are respectively devoted to the study of solitons with power law, parabolic law, dual-power law, and saturable law nonlinearities. In each case, we have developed the soliton dynamics, evaluated integrals of motion, and devoted enough space to develop adiabatic dynamics of perturbed quantities based on multiple-scale perturbation theory. In addition, the existence of bistable soliton is discussed in chapter 7. Chapter 8 is devoted to intrachannel collision of optical solitons in the presence of perturbation terms. Both Hamiltonian as well as non-Hamiltonian type perturbations have been considered. The nonlinearities that are studied in this chapter are Kerr, power, parabolic, and dual-power laws. In chapter 9, the stochastic perturbation of optical solitons has been studied. The corresponding Langevin equations are derived and analyzed for each of the laws of nonlinearity, namely Kerr, power, parabolic, and dual-power laws. Optical couplers are introduced in chapter 10. Twin core and multiple-core couplers have been discussed. At the end of this chapter, we have briefly discussed solitons in magneto-optic waveguides. The book concludes with chapter 11, which treats an introduction to optical bullets.

This book is intended for graduate students at the master's and doctoral levels in applied mathematics, physics, and engineering. Undergraduate students with senior standing in applied mathematics, physics, and engineering will also benefit from this book. The prerequisite of this book is a knowledge of partial differential equations, perturbation theory, and elementary physics.

Anjan Biswas, is extremely thankful to Dr. Michael Busby, director of the Center of Excellence in Information Systems Engineering and Management of Tennessee State University in Nashville, Tennessee, with which this author was previously affiliated. Without constant encouragement and financial support from Dr. Busby, this project would not have been possible. The first author is also extremely thankful and grateful to Dr. Tommy Frederick, vice provost of research at Delaware State University, with which this author is presently affiliated, for his constant encouragement. Without these two persons' blessings, this project would not have been possible. Finally, the author is extremely grateful to his parents for all their unconditional love in his upbringing, blessings, education, support, encouragement, and sacrifices throughout his life, till today. This author is deeply saddened by the sudden death of his mother after a massive heart attack in Calcutta, India, which occurred during the course of writing this book.

Swapan Konar is grateful to Prof. H. C. Pande, vice chancellor emeritus, Birla Institute of Technology, India; Prof. S. K. Mukherjee, vice chancellor, Birla Institute of Technology, and Prof. P. K. Barhai, head of the Department of Applied Physics, Birla Institute of Technology, for encouragement and constant support. Finally, he sincerely thanks his wife Tapati for her tolerance and encouragement and his little son Argho for sacrificing his playtime.

Authors

Dr. Anjan Biswas obtained his BSc (Honors) in mathematics from St. Xavier's College, Calcutta, and subsequently earned his MSc and MPhil degrees in applied mathematics from the University of Calcutta, India. After that, he obtained his MA and PhD degrees in applied mathematics from the University of New Mexico, Albuquerque. His current research interests are in nonlinear optics, theory of solitons, plasma physics, and fluid dynamics. He is the author of 100 refereed journal papers and also serves as an editorial board member for three journals. Currently, he is an associate professor in the Department of Applied Mathematics and Theoretical Physics of Delaware State University in Dover.

Dr. S. Konar received his MSc degree in Nuclear Physics from University of Kalyani, India, in 1982. He earned an MTech in energy management. Dr. Konar has been awarded MPhil and PhD respectively in 1987 and 1990 by Jawaharlal Nehru University, New Delhi, India. At present, he is working as a professor of applied physics at Birla Institute of Technology, Mesra, Ranchi, India. His current research interest is in the field of photonics and optoelectronics, particularly classical solitons, soliton propagation in dispersion-managed optical communication systems, nonlinear optical waveguide, induced focusing, self-focusing and all optical switching. He has published 64 research papers in international journals and presented 30 research papers in conferences.

Contents

1

Introduction

This introductory chapter is intended to provide a general overview of fiber optics. It starts with a history of and current developments in fiber optics in Section 1.1. Section 1.2 provides a brief account of types of optical waveguides and the issues of fiber-optic communications.

1.1 History

The propagation of optical pulses, or solitons, through optical fibers has been a major area of study given its potential applicability in optical communication systems. The field of telecommunications has undergone a substantial evolution in the last couple of decades due to the impressive progress in the development of optical fibers, optical amplifiers, and transmitters and receivers. In a modern optical communication system, the transmission link is composed of optical fibers and amplifiers that replace the electrical regenerators. However, the amplifiers introduce some noise and signal distortion that limit the system capacity. Presently, the optical systems that show the best characteristics in terms of simplicity, cost, and robustness against the degrading effects of a link are those based on intensity modulation with direct detection (IM-DD). Conventional IM-DD systems are based on the non-return-to-zero (NRZ) format, but for soliton-based transmission at higher data rates the return-to-zero (RZ) format is used. Soliton-based transmission allows the exploitation of the fiber capacity much more. [9].

The performance of optical system is limited by several effects that are present in optical fibers and amplifiers. Signal propagation through optical fibers can be affected by group velocity dispersion (GVD), polarization mode dispersion (PMD), and nonlinear effects. The chromatic dispersion that is essentially the GVD when waveguide dispersion is negligible is a linear effect that introduces pulse broadening and generates intersymbol interference. The PMD arises due to the fact that optical fibers for telecommunications have two polarization modes, in spite of the fact that they are called *monomode fibers*. These modes have two different group velocities that induce pulse broadening depending on the input signal state of polarization. The transmission impairment due to PMD looks similar to that caused by GVD. However, PMD is random whereas GVD is a deterministic process. So, PMD cannot be controlled at the receiver. Newly installed optical fibers have quite low values of PMD that are about 0.1 $ps/\sqrt{km}$.

The main nonlinear effects that arise in monomode fibers are Brillouin scattering, Raman scattering, and the Kerr effect. Brillouin is a backward scattering that arises from acoustic waves and can generate noise at the receiver. Raman scattering takes place in both forward and backward directions from silica molecules. The Raman gain response is characterized by low gain and wide bandwidth, namely about 30 THz. The Raman threshold in conventional fibers is of the order of 500 mW for copolarized pump and Stokes' wave (that is about 1 W for random polarization), thus making Raman effect negligible for a single-channel signal. However, it becomes important for multichannel wavelength-division-multiplexed (WDM) signals due to an extremely wide band of wide gain curve.

The Kerr effect of nonlinearity is due to the dependence of the fiber refractive index on the field intensity. The intensity dependence of the refractive index leads to a larger number of interesting nonlinear effects. Notable among them, which have been studied widely, are self-phase modulation (SPM) and cross-phase modulation (XPM). SPM refers to the self-induced nonlinear phase shift experienced by an optical field during its propagation through an optical fiber. SPM is responsible for spectral broadening. The SPM-induced chirp combines with the linear chirp generated by the chromatic dispersion. If the fiber dispersion coefficient is positive, namely in the normal dispersion regime, linear and nonlinear chirps have the same sign, whereas in an anomalous dispersion regime, they are of opposite signs. In the former case, pulse broadening is enhanced by SPM, while in the later case it is reduced. In the anomalous dispersion case, the Kerr nonlinearity induces a chirp that can compensate the degradation induced by GVD. Such a compensation is total if soliton signals are used.

If multichannel WDM signals are considered, the Kerr effect can be more degrading since it induces nonlinear cross-talk among the channels that are known as XPM. In addition, WDM generates new frequencies called four-wave mixing (FWM). The other issue in the WDM system is the collision-induced timing jitter that is introduced due to the collision of solitons in different channels. The XPM causes further nonlinear chirp that interacts with the fiber GVD as in the case of SPM. The FWM is a parametric interaction among waves that satisfies a particular relationship called *phase-matching* that leads to power transfer among different channels.

To limit the FWM effect in a WDM, it is preferable to operate with a local high GVD that is periodically compensated by devices having an opposite GVD sign. One such device is a simple optical fiber with appropriate GVD, and the method is commonly known as *dispersion management*. With this approach, the accumulated GVD can be very low and, at the same time, FWM is strongly limited. Through dispersion management it is possible to achieve the highest capacity for both RZ and NRZ signals. In that case, the overall link dispersion has to be kept very close to zero, while a small amount of chromatic anomalous dispersion is useful for the efficient propagation of a soliton signal. It has been demonstrated with soliton signals that dispersion management is very useful since it reduces collision-induced timing jitter and also pulse interactions. It thus permits the achievement of higher capacities than those allowed by the link having constant chromatic dispersion.

1.2 Optical Waveguides

One of the most promising applications of soliton theory is in the field of optical communications. In optical communications systems, information is encoded into light pulses and transmitted through optical fibers over long distances. Commercial systems have been in operation since 1977 and a transatlantic undersea optical cable has been developed. In 1973, Hasegawa and Tappert [183, 184] proposed that soliton pulses could be used in optical communications. However, the technology was not available until 7 years later, at which time researchers at Bell Laboratories had experimentally demonstrated the propagation of solitons in optical fibers.

Rapid developments in communications technology have occurred— for example, the change from the use of wires to send signals (wire telegraphy) to wireless or radio telegraphy—leading to enormously increased communication rates, measured by bits per second by a factor of 1 billion. The latest in this series of advances is the optical fiber system in which large amounts of information, coded as light pulses, pass along silica fibers. The first transoceanic links, namely along the Atlantic and Pacific oceans, have been established. As marvelous as these advances have been, the present system still uses only a tiny fraction of the information-carrying capacity of optical fibers.

Taking a look at waveguides, in particular an optical fiber, one can see how solitons promise to revolutionize the field of telecommunications. The main idea of a waveguide is to guide a beam of light by employing a variation of refractive index in the transverse direction so as to cause the light to travel along a well-defined channel. The dependence of refractive index on the transverse direction, the direction perpendicular to that in which the wave propagates, can be continous or discontinous. The essential feature, however, is that the refractive index is maximal in the channel along which one wishes the light to be guided.

Figures 1.1(a) and 1.1(b) show the cross-section of an optical fiber. The inner core consists of a special form of silica glass with very low absorption and is between 10 to 60 μm in diameter. This core is surrounded by a glass cladding whose refractive index, n_2, is very close to but slightly less than n_1, the linear refractive index of the inner core. This ensures that the wave is

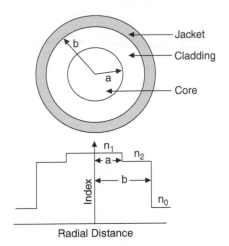

FIGURE 1.1
(a) & (b) Cross section of an optical fiber.

guided, namely its intensity is largely confined, to the inner core by virtue of total internal reflection.

Two parameters characterize an optical fiber, namely the core-cladding index difference (Δ) that is defined as:

$$\Delta = \frac{n_1 - n_2}{n_1} \tag{1.1}$$

and the normalized frequency (V) that is defined as:

$$V = \frac{2\pi a}{\lambda}\sqrt{n_1^2 - n_2^2} \tag{1.2}$$

where a is the radius of the fiber core as shown in Figure 1.2 and λ is the wavelength of light. The parameter V determines the number of modes supported by the fiber. For a V less than 2.405 the fibers support a single mode and so the fibers that are designed to satisfy such conditions are known as

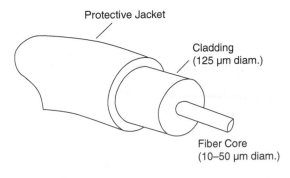

FIGURE 1.2
Structure of an optical fiber.

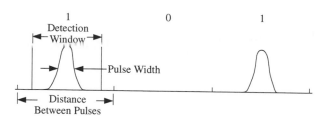

FIGURE 1.3
Soliton detection window and soliton train.

single-mode fibers. A typical multimode fiber would have the core radius as $a = 25$–30 μm. However, for a single-mode fiber, a typical value of Δ is $\sim 3 \times 10^{-3}$ and requires a to be in the range of 2–4 μm. The value of the outer radius b is less critical as long as it is large enough to confine the modes entirely. Typically, $b = 50$–60 μm for both single-mode and multimode fibers (Figure 1.2).

The basic idea of using optical fibers for communications is relatively simple. The message is coded in binary by representing one as a pulse-like modulation of a carrier wave whose wavelength is in the micrometer (10^{-6} m) range and whose frequency is in the terahertz (10^{14} Hz) range and representing zero by the absence of such a pulse. The arrangement is shown in Figure 1.3. The pulses are approximately 10–25 picoseconds (10^{-12} s) wide and the average distance between them is four times that amount. Experimentally, fibers have managed effective transmission rates in the gigabit range (10^9 bits/s).

1.2.1 Types of Optical Fibers

Based on the refractive index profile there are two types of optical fibers:

1. *Step index fiber*: In a step index fiber, the refractive index of the core is uniform throughout and undergoes an abrupt or a step change at the core-cladding boundary.
2. *Graded index fiber*: In a graded index fiber, the refractive index of the core is made to vary in a parabolic manner such that the maximum value of the refractive index is at the center of the core (Figure 1.4).

Propagating rays in the fiber can be classified as meridional and skew rays. Meridional rays are confined to the meridional plane of the fiber, which are planes that contain the axis of symmetry of the fiber. Skew rays are not confined to a single plane. They propagate along the fiber.

1.2.2 Advantages of Fiber-Optic Communications

The various advantages of soliton communication through optical fibers are enumerated here:

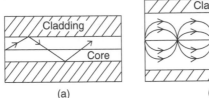

FIGURE 1.4
(a) Step-index fiber and (b) graded-index fiber.

1. *Wider bandwidth:* The information-carrying capacity of a transmission system is directly proportional to the carrier frequency of the transmitted signals. The optical carrier frequency is in the range of $10^{13}-10^{15}$ Hz while the radio wave frequency is about 10^6 Hz and the microwave frequency is about 10^{10} Hz. Thus, the optical fiber yields greater transmission bandwidth than conventional communication systems and the data rate or number of bits per second is increased to a greater extent in the optical fiber communication system.

2. *Low transmission loss:* Due to the usage of ultra-low-loss fibers and erbium-doped silica fibers as optical amplifiers, one can achieve almost lossless transmission. In modern optical fiber telecommunication systems, the fibers having a transmission loss of 0.2 dB/km are used. Furthermore, using erbium-doped silica fibers over a short length in transmission path selective points, appropiate optical amplification can be achieved. Thus, the repeater spacing is more than 100 km. Since the amplification is done in the optical domain itself, the distortion produced during the strengthening of the signal is almost negligible.

3. *Dielectric waveguide:* Optical fibers are made from silica, which is an electrical insulator. Therefore, they do not pick up any electromagnetic waves or any high-current lightning. They are also suitable in explosive environments. Furthermore, the optical fibers are not affected by any interference originating from power cables, railway power lines, and radio waves. There is no cross-talk between the fibers in a cable because of the absence of optical interference between the fibers.

4. *Signal security:* Signals transmitted through the fibers do not radiate. In addition, signals cannot be tapped from a fiber in an easy manner. Therefore, optical communication provides 100% signal security.

5. *Small size and weight:* Fiber-optic cables are developed with small radii and are flexible, compact, and lightweight. They can be bent or twisted without any damage. Optical fiber cables are superior to copper cables in terms of storage, handling, installation, and transportation, maintaining comparable strength and durability.

2

The Nonlinear Schrödinger's Equation

This chapter will talk about the mathematical aspects of the nonlinear Schrödinger's equation (NLSE) that governs the propagation of solitons through an optical fiber. Section 2.1 is an introduction to NLSE. In Section 2.2, the conserved quantities of the NLSE will be derived. In Section 2.3 the soliton parameters will be introduced and the formulae for the adiabatic dynamics of these parameters in the presence of the perturbation terms will be given. Finally, in Section 2.4, the concept of quasi-stationarity will be introduced.

2.1 Introduction

The NLSE plays a vital role in various areas of physical, biological, and engineering sciences. It appears in many applied fields, including fluid dynamics, nonlinear optics, plasma physics, and protein chemistry. The NLSE that is going to be studied in this book is given by [86, 108, 399]

$$iq_t + \frac{1}{2}q_{xx} + F(|q|^2)q = 0 \tag{2.1}$$

In (2.1), F is a real-valued algebraic function and one needs to have the smoothness of the complex function $F(|q|^2)q : C \mapsto C$. Considering the complex plane C as a two-dimensional linear space R^2, it can be said that the function $F(|q|^2)q$ is k times continuously differentiable so that one can write

$$F(|q|^2)q \in \bigcup_{m,n=1}^{\infty} C^k((-n, n) \times (-m, m); R^2)$$

In equation (2.1), q is the dependent variable, x and t are the independent variables, and the subscripts represent the partial derivative of q with respect to that variable. So, q_t stands for $\partial q/\partial t$ while q_{xx} stands for $\partial^2 q/\partial x^2$. The first term in (2.1) represents the time evolution term, while the second term is due to the group velocity dispersion and the third term accounts for nonlinearity. Thus, equations of these types are sometimes known as *nonlinear evolution*

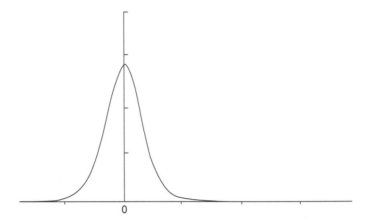

FIGURE 2.1
Profile of a soliton.

equations. This is a nonlinear partial differential equation that is not integrable, in general. The nonintegrability is not necessarily related to the nonlinear term in (2.1). Higher order dispersion, for example, can also make the system nonintegrable while it still remains Hamiltonian.

Equation (2.1) has been shown to govern the evolution of a wave packet in a weakly nonlinear and dispersive medium and has thus arisen insuch diverse fields as water waves, plasma physics, and nonlinear optics. One other application of this equation is in pattern formation, where it has been used to model some nonequilibrium pattern forming systems. In particular, this equation is now widely used in the optics field as a good model for optical pulse propagation in nonlinear fibers. Equation (2.1) is known to support solitons or soliton solutions for various kinds of nonlinearity that will be discussed in the upcoming chapters. The term *soliton* refers to a nonlinear wave that propagates without changing properties and is stable against mutual collisions with other solitons that retain their identities. Solitons have been studied extensively in various areas of mathematical physics. In the context of optical fibers, solitons are not only of fundamental interest but also have potential applications in the field of optical fiber communications. Figure 2.1 shows an illustration of a soliton. This text is devoted to the study of the propagation of such solitons through optical fibers with emphasis on the various kinds of the function $F(s)$ in equation (2.1).

2.1.1 Nonlinearity Classification

There are various kinds of nonlinearities of the function F in (2.1) that are known so far. They are as follows:

1. *Kerr law*: $F(s) = s$
   This is also known as *cubic nonlinearity* and is the simplest known form of the law of nonlinearity. In this case, the NLSE is integrable by a method called the Inverse Scattering Transform. This method will

be discussed in the next chapter. Most optical fibers that are commercially available nowadays obey this Kerr law of nonlinearity.

2. *Power law* $F(s) = s^p$

   In this case, it is necessary to have $0 < p < 2$ to avoid wave collapse. In fact, it is mandatory that $p \neq 2$ to avoid self-focusing singularity. This law of nonlinearity arises in nonlinear plasmas and solves the problem of small K-condensation in weak turbulence theory. It also arises in the context of nonlinear optics. Physically, various materials, including semiconductors, exhibit power law nonlinearities. This case of nonlinearity has been studied, including the perturbation term by multiple-scale analysis. The case where $p = \frac{1}{2}$ is studied in the context of soliton turbulence.

3. *Parabolic law*: $F(s) = s + \nu s^2$

   This law is commonly known as the *cubic-quintic nonlinearity*. The second term is large for the case of p-toluene sulfonate crystals. This law arises in the nonlinear interaction between Langmuir waves and electrons. It describes the nonlinear interaction between the high-frequency Langmuir waves and the ion-acoustic waves by pondermotive forces. This case of cubic-quintic nonlinearity was also studied by multiple-scale analysis.

4. *Dual-power law*: $F(s) = s^p + \xi s^{2p}$

   This model is used to describe the saturation of the nonlinear refractive index, and its exact soliton solutions are known. The effective GNLSE with this dual-power law nonlinearity serves as a basic model to describe spatial solitons in photovoltaic-photorefractive materials such as lithium niobate. Optical nonlinearities in many organic and polymer materials can be modelled using this form of nonlinearity. The solitons of this model become unstable and decay in the unstable region $1 \leq p < 2$, while for $p \geq 2$, the solitons collapse in a finite time.

5. *Saturating law*: $F(s) = \frac{\lambda s}{1+\lambda s}$

   This law with $\lambda > 0$ accurately describes the variation of the dielectric constant of gas vapors through which a laser beam propagates [30]. Optical nonlinearity saturates at a finite value of optical intensity in most materials. $F(s)$ in those materials can be modeled using the above form, which is known as the *saturating form* of nonlinearity. In semiconductor-doped fibers, the soliton propagation has been modeled using saturable nonlinearity rather than the usual Kerr nonlinearity. The main motivation behind such attempts is the observation of such nonlinearity at not too high intensities in semiconductor-doped glass and other composite materials. This case was studied numerically.

6. *Exponential law*: $F(s) = \frac{1}{2\lambda}(1 - e^{-2\lambda s})$

   This case of exponential nonlinearity serves as useful model in homogenous, unmagnetized plasmas and laser-produced plasmas.

When the phase velocity of slow plasma oscillation is much smaller than the ion thermal velocity, one can obtain the adiabatic or quasi-static electron density under the quasi-neutral approximation. Now, combining the coupling equation that exhibits the slowly varying complex amplitude with the low frequency plasma motion, one obtains the saturable law of nonlinearity.

7. *Log law*: $F(s) = a \ln(b^2 s)$
   This law arises in various fields of contemporary physics. It allows closed form exact expressions for stationary Gaussian beams (gaussons) as well as for periodic and quasi-periodic regimes of beam evolution. The advantage of this model is that the radiation from the periodic soliton is absent because the linearized problem has a discrete spectrum only.

8. *Higher order polynomial law*: $F(s) = s + \nu s^2 + \gamma s^3$
   This law is an extension of the parabolic law that is given in #3. The Hamiltonian energy diagrams are studied in [35]. This law is also observed in various physical systems [31, 35].

9. *Triple-power law*: $F(s) = s^p + \nu s^{2p} + \gamma s^{3p}$
   This law is an extension of the dual-power law and is a generalization of the higher order polynomial law. In this case, the Hamiltonian energy diagrams are also studied.

10. *Threshold law*:

$$F(s) = \begin{cases} n_1 & : \quad s < I_0 \\ n_2 & : \quad s \geq I_0 \end{cases}$$

A smooth transition of this kind can be modeled as

$$F(s) = As\{1 + \alpha \tanh[\gamma(s^2 - I_0^2)]\}$$

where for $s \ll I_0$, $F(s) \approx n_1 s$, where $n_1 = A[1 - \alpha \tanh^2(\gamma I_0^2)]$, and for $s \gg I_0$, $F(s) \approx n_2 s$, where $n_2 = 1 + \alpha$. Although examples of nonlinear optical materials with this law are not yet known, the bistable solitons have interesting properties that can be useful for future applications in all-optical logic and switching devices.

Of these ten forms of the function F, the first five laws of nonlinearity will be studied in this text in the upcoming chapters. Of these first five laws, the first four forms permit exact soliton solutions; however, for the saturable law, an exact soliton solution is not known and so this form of nonlinearity will be studied by means of the variational principle that will be introduced in the next chapter.

To date, no closed-form soliton solutions are known besides the first four forms of nonlinearity. However, a considerable amount of research is on going with higher order polynomial law and triples-power law nonlinearity to obtain the closed form of optical soliton solution.

2.2 Travelling Waves

A traveling wave is a solution of the NLSE that represents a wave of permanent form that does not change its shape during propagation and moves with a constant speed. The wave may be localized or periodic. In order to seek a traveling wave solution of the NLSE that is given by (2.1), introduce the ansatz

$$q(x, t) = Ag[B(x - \bar{x})]e^{i(-\kappa x + \omega t + \sigma_0)} \tag{2.2}$$

where the function g represents the shape of the soliton described by (2.1) and depends on the type of nonlinearity in it. Also, in (2.2) A and B respectively represent the amplitude and width of the soliton, κ is the soliton frequency, ω is the soliton wave number, σ_0 is the center of phase of the soliton, and $\bar{x}$ gives the mean position of the soliton so that the velocity of the soliton is given by

$$v = \frac{d\bar{x}}{dt} \tag{2.3}$$

The soliton width and amplitude are related as $B = \Lambda(A)$, where the functional form λ depends on the type of nonlinearity in (2.1). Thus, from (2.2), one easily gets

$$q_t = -ABvg'(\tau)e^{i\phi} - i\omega Ag(\tau)e^{i\phi} \tag{2.4}$$

where, the phase ϕ is

$$\phi(x, t) = -\kappa x + \omega t + \sigma_o \tag{2.5}$$

and

$$\tau = B(x - \bar{x}) \tag{2.6}$$

Also, from (2.2) one can get

$$q_{xx} = AB^2 g''(\tau) - 2i\kappa ABg'(\tau) - \kappa^2 Ag(\tau) \tag{2.7}$$

Substituting (2.4) and (2.7) into (2.1) simplifies it to

$$-iBvg' + \omega g + \frac{1}{2}B^2 g'' - i\kappa Bg' - 1/2\kappa^2 g + gF(A^2 g^2) = 0 \tag{2.8}$$

From (2.8), equating the real and imaginary parts yields

$$\Lambda Bg'(\kappa + v) = 0 \tag{2.9}$$

and

$$B^2 g'' - (\kappa^2 - 2\omega)g + 2gF(A^2 g^2) = 0 \tag{2.10}$$

so that from (2.9), one obtains

$$\kappa = -v \tag{2.11}$$

Now, multiplying both sides of (2.10) by g', then integrating and choosing the integration constant to be zero since the wave profile is such that q, q_x, and q_{xx} approach zero as $|x| \to \infty$ gives

$$B^2(g')^2 - (\kappa^2 - 2\omega)g^2 + 2\int (g^2)' F(A^2 g^2) dg = 0 \tag{2.12}$$

On separation of variables and integration once more, equation (2.12) leads to

$$x - vt = \int \frac{dg}{[(\kappa^2 - 2\omega)g^2 - 2\int (g^2)' F(A^2 g^2)]^{\frac{1}{2}}} \tag{2.13}$$

Equation (2.13) can be integrated only if the law of nonlinearity is known. As discussed in subsequent chapters, this equation is possible to integrate for the cases of Kerr law, power law, parabolic law, and dual-power law only. In all other cases of nonlinearity, the closed-form soliton solution is not yet known and remains to be established. This equation will be integrated completely in the subsequent chapters and the solitons corresponding to the respective laws of nonlinearity will be studied in detail.

2.3   Integrals of Motion

An important property of nonlinear evolution equations is that they have conserved quantities known as *integrals of motion*. In this section, the conserved quantities of the NLSE, given by (2.1), will be derived. Conservation laws are a common feature in mathematical physics, where they describe the conservation of fundamental physical quantities. Rewriting the NLSE in the form [137]

$$\frac{\partial T}{\partial t} + \frac{\partial X}{\partial x} = 0 \tag{2.14}$$

represents the *conservation law*. Here T is known as the *density* while X is known as the *flux*. Neither the density nor the flux involve derivatives with respect to t. Thus T and X may depend upon $x, t, q, q_x, q_{xx}, \ldots$, but not q_t. Now, if both T and X_x are integrable on $(-\infty, \infty)$, so that $X \longrightarrow$ constant as $|x| \longrightarrow \infty$, then equation (2.1) can be integrated to yield

$$\frac{d}{dt}\left(\int_{-\infty}^{\infty} T dx\right) = 0 \tag{2.15}$$

so that

$$\int_{-\infty}^{\infty} T \, dx = constant \tag{2.16}$$

The integral of T, over all x, is therefore called the *constant of motion* or the *integral of motion*. For a dynamical system with a finite number of degrees of freedom to be integrable, the system needs to have as many conserved quantities as the degrees of freedom. The first conserved quantity for the NLSE will now be derived.

Performing the operation (2.1) $\times q^*$ yields

$$iq^*q_t + \frac{1}{2}q^*q_{xx} + F(|q|^2)|q|^2 = 0 \tag{2.17}$$

The complex conjugate of equation (2.17) is

$$-iqq_t^* + \frac{1}{2}qq_{xx}^* + F(|q|^2)|q|^2 = 0 \tag{2.18}$$

Operating (2.17)–(2.18) gives

$$i(q^*q_t + qq_t^*) + \frac{1}{2}(q^*q_{xx} - qq_{xx}^*) = 0 \tag{2.19}$$

which can be rewritten as

$$i(|q|^2)_t + \frac{1}{2}(q^*q_x - qq_x^*)_x = 0 \tag{2.20}$$

so that the flux is $q^*q_x - qq_x^*$. Integrating equation (2.20) with respect to x yields

$$\frac{d}{dt}\left(\int_{-\infty}^{\infty} |q|^2 dx\right) = 0 \tag{2.21}$$

since, for a soliton, $q, q_x, q_{xx}, \ldots$ approach zero as $|x| \longrightarrow \infty$, as previously mentioned. Thus, the first conserved quantity for the NLSE is given by

$$E = \int_{-\infty}^{\infty} |q|^2 dx = constant \tag{2.22}$$

This conserved quantity is known as the *wave energy* or the *mass, wave action*, or *plasmon number*; in optics, however, it is called the *wave power*, while mathematically, it is known as the L_2 *norm*.

Now the second conserved quantity will be derived. The complex conjugate of (2.1) is given by

$$-iq_t^* + \frac{1}{2}q_{xx}^* + F(|q|^2)q^* = 0 \tag{2.23}$$

Performing the operation $q_x^* \times (2.1) + q_x \times (2.23)$ gives

$$i(q_t q_x^* - q_x q_t^*) + \frac{1}{2}(q_x^* q_{xx} + q_x q_{xx}^*) + (q q_x^* + q^* q_x)F(|q|^2) = 0 \quad (2.24)$$

Now performing the operation $q^* \times \frac{\partial}{\partial x}(2.1) + q \times \frac{\partial}{\partial x}(2.23)$ gives

$$i(q^* q_{xt} - q q_{xt}^*) + \frac{1}{2}(q^* q_{xxx} + q q_{xxx}^*)$$

$$+ \left[q^* \frac{\partial}{\partial x}\{F(|q|^2)q\} + q\frac{\partial}{\partial x}\{F(|q|^2)q^*\} \right] = 0 \quad (2.25)$$

Then (2.24)–(2.25) gives

$$i\frac{\partial}{\partial t}(q q_x^* - q^* q_x) + \frac{1}{2}\frac{\partial}{\partial x}(q_x q_x^*) - \frac{1}{2}(q^* q_{xxx} + q q_{xxx}^*)$$

$$+ (|q|^2)_x F(|q|^2) - [(|q|^2)_x F(|q|^2) + 2|q|^2 F'(|q|^2)] = 0 \quad (2.26)$$

Equation (2.26) can be rewritten as

$$i\frac{\partial}{\partial t}(q q_x^* - q^* q_x) + \frac{1}{2}\frac{\partial}{\partial x}(q_x q_x^*)$$

$$- \frac{1}{2}\frac{\partial}{\partial x}(q^* q_{xx} + q q_{xx}^*) + \frac{1}{2}\frac{\partial}{\partial x}(q_x q_x^*) - 2|q|^2 F'(|q|^2) = 0 \quad (2.27)$$

which simplifies to

$$i\frac{\partial}{\partial t}(q q_x^* - q^* q_x) + \frac{1}{2}\frac{\partial}{\partial x}[2|q_x|^2 - (q^* q_{xx} + q q_{xx}^*) - 4F(|q|^2)] = 0 \quad (2.28)$$

Integrating (2.28) with respect to x gives

$$i\frac{d}{dt}\int_{-\infty}^{\infty}(q^* q_x - q q_x^*)dx = 0 \quad (2.29)$$

so that

$$M = i\int_{-\infty}^{\infty}(q^* q_x - q q_x^*)dx = constant \quad (2.30)$$

which is the second conserved quantity, also known as the *linear momentum*. Equation (2.1) also has a third conserved quantity that is given by [86, 90]

$$H = \int_{-\infty}^{\infty}\left[\frac{1}{2}|q_x|^2 - f(I)\right]dx \quad (2.31)$$

where

$$f(I) = \int_0^I F(\xi)d\xi \quad (2.32)$$

with the intensity I as $I = |q|^2$. This integral of motion is known as the *Hamiltonian*. The derivation of this conserved quantity is left to the reader as an exercise. Note that, for the energy conservation law, there are no contributions from the nonlinear and the dispersion terms of the NLSE. Also, for the conservation of linear momentum, there are no contributions from the evolution term and the nonlinear term. Finally, for the Hamiltonian conservation, there is no contribution from the evolution term.

The Hamiltonian is one of the most fundamental notions in mechanics and, more generally, in the theory of conservative dynamical systems with finite or even infinite degrees of freedom. The Hamiltonian formalism has turned out to be one of the most universal in the theory of integrable system and nonlinear waves in general. In the case of nonintegrable systems, the Hamiltonian exists whenever the system is conservative and it is useful for stability analysis. The most useful approach in the soliton theory of conservative nonintegrable Hamiltonian system is a representation on the plane of conserved quantities, namely the Hamiltonian-versus-energy diagrams [35]. In a two-parameter family of solitons, the Hamiltonian-momentum-energy diagrams are quite informative.

The Hamiltonian-versus-energy diagrams have been effectively used to study the various properties of solitons, namely their range of existence, stability, and general dynamics. These properties were studied in the context of scalar non–Kerr law solitons, vector solitons in birefringent waveguides, radiation phenomena from unstable soliton branches, optical couplers, the theory of Bose-Einstein condensates, and many more. These Hamiltonian-versus-energy curves are also useful for analyzing the stability of bound states, whose definition is introduced in chapter 3.

One can see that (2.1) can now be written in canonical forms

$$iq_t = \frac{\delta H}{\delta q^*} \tag{2.33}$$

$$iq_t^* = -\frac{\delta H}{\delta q} \tag{2.34}$$

where in (2.33) and (2.34) the right sides denote the Fréchet derivative $\delta F / \delta q$ that is defined as

$$\int_{-\infty}^{\infty} v \frac{\delta F}{\delta q} dx = \lim_{\epsilon \to 0} \int_{-\infty}^{\infty} F(q + \epsilon v) dx \tag{2.35}$$

for all continuous v. Thus (2.33) and (2.34) define a Hamiltonian dynamical system on an infinite dimensional phase space. This system can be analyzed using the theory of Hamiltonian systems that the behavior of the solution is defined, to a large extent, by the singular points of the system, namely the stationary solutions of (2.1), and depends on the nature of these points as determined by the stability of its stationary solutions.

The soliton solution of (2.1), although not integrable, is assumed to be given in the form, as introduced in the traveling wave ansatz in (2.2)

$$q(x, t) = A(t)g[B(t)\{x - \bar{x}(t)\}]e^{i\phi(x,t)} \tag{2.36}$$

where $\phi(x, t)$ is the phase of the soliton that is given by

$$\phi(x, t) = -\kappa x + \omega t + \sigma_0 \tag{2.37}$$

so that

$$\frac{\partial \phi}{\partial x} = -\kappa = v \tag{2.38}$$

and

$$\frac{\partial \phi}{\partial t} = \omega = \frac{B^2}{2} \frac{I_{0,0,2,0}}{I_{0,2,0,0}} - \frac{\kappa^2}{2} + \frac{1}{I_{0,2,0,0}} \int_{-\infty}^{\infty} g^2(s) F(A^2 g^2(s)) ds \tag{2.39}$$

where

$$I_{\alpha,\beta,\gamma,\nu} = \int_{-\infty}^{\infty} \tau^\alpha g^\beta(\tau) \left(\frac{dg}{d\tau}\right)^\gamma \left(\frac{d^2 g}{d\tau^2}\right)^\nu d\tau \tag{2.40}$$

for nonnegative integers α, β, γ, and ν with $\tau = B(t)(x - \bar{x}(t))$. Also, the velocity ($v$) of the soliton is defined in (2.3). Here, (2.39) is obtained by differentiating (2.36) with respect to t and subtracting from its conjugate while using (2.1). For such a general form of the soliton given by (2.36), the integrals of motion, from (2.22), (2.30), and (2.31) respectively, reduce to

$$E = \int_{-\infty}^{\infty} |q|^2 dx = \frac{A^2}{B} I_{0,2,0,0} \tag{2.41}$$

$$M = \frac{i}{2} \int_{-\infty}^{\infty} (qq_x^* - q^* q_x) dx = -\kappa \frac{A^2}{B} I_{0,2,0,0} \tag{2.42}$$

$$H = \int_{-\infty}^{\infty} \left[\frac{1}{2}|q_x|^2 - f(|q|^2)\right] dx$$

$$= \frac{A^2 B}{2} I_{0,0,2,0} + \frac{\kappa^2 A^2}{2B} I_{0,2,0,0} - \int_{-\infty}^{\infty} \int_0^I F(s) ds dx \tag{2.43}$$

The exact forms of these conserved quantities will be derived in the following chapters for the various forms of nonlinearity that are going to be considered.

2.4 Parameter Evolution

The parameters for the soliton given by (2.36) are now defined as follows [86, 185, 191]

$$\kappa(t) = -\frac{\partial\phi}{\partial x} = \frac{i}{2}\frac{\int_{-\infty}^{\infty}(qq_x^* - q^*q_x)dx}{\int_{-\infty}^{\infty}|q|^2dx} \tag{2.44}$$

$$\bar{x}(t) = \frac{\int_{-\infty}^{\infty}x|q|^2dx}{\int_{-\infty}^{\infty}|q|^2\,dx} = \frac{1}{E}\int_{-\infty}^{\infty}x|q|^2dx \tag{2.45}$$

where E in (2.45) represents the energy given by (2.41). To obtain the evolution of the soliton parameters, differentiate $\kappa(t)$ and $\bar{x}(t)$ given by (2.44) and (2.45) respectively with respect to t, keeping in mind that E is a constant of motion. This leads to the following evolution equations for the soliton frequency

$$\frac{d\kappa}{dt} = 0 \tag{2.46}$$

$$\frac{d\bar{x}}{dt} = -\kappa \tag{2.47}$$

The velocity of the soliton is obtained. Again, the fact that E is a conserved quantity and $B(t) = \lambda(A(t))$ are true leads to the result

$$\frac{dA}{dt} = \frac{dB}{dt} = 0 \tag{2.48}$$

from (2.41) so that the soliton amplitude and width stay constant during propagation.

2.4.1 Perturbation Terms

In this book, the NLSE is going to be considered along with its perturbation terms. Perturbation terms do arise in the context of optics and cannot be avoided, as will be seen. The NLSE, along with its perturbation terms, is given by

$$iq_t + \frac{1}{2}q_{xx} + F(|q|^2)q = i\epsilon R[q, q^*] \tag{2.49}$$

where R is a spatio-differential operator while the perturbation parameter ϵ with $0 < \epsilon \ll 1$ is called the relative width of the spectrum that arises due to quasi-monochromaticity [145] in nonlinear fiber optics, namely [185]

$$\epsilon = \frac{\Delta\omega_0}{\omega_0} = \frac{\omega - \omega_0}{\omega_0} \tag{2.50}$$

In (2.50), ω_0 is the carrier frequency of the light wave, while $\Delta\omega_0$ represents the departure from the carrier frequency. By quasi-monochromaticity, one means that the spectrum centered at ω_0 has a spectral width $\Delta\omega$ such that $\Delta\omega/\omega_0 \ll 1$. Since $\omega_0 \sim 10^{15}s^{-1}$, quasi-monochromaticity is valid for pulses whose widths are ≥ 0.1 ps ($\Delta\omega \leq 10^{13}s^{-1}$).

In the presence of perturbations, the conserved quantities no longer exist. Instead, they are *modified integrals of motion*. Now differentiate $E(t)$, $M(t)$, and $H(t)$ that are respectively given by equations (2.41), (2.42), and (2.43) using (2.49). The following adiabatic evolution equations for the integrals of motion are obtained

$$\frac{dE}{dt} = \epsilon \int_{-\infty}^{\infty} (q^*R + qR^*)dx \tag{2.51}$$

$$\frac{dM}{dt} = i\epsilon \int_{-\infty}^{\infty} (q_x^*R - q_xR^*)dx \tag{2.52}$$

$$\frac{dH}{dt} = 2\epsilon \int_{-\infty}^{\infty} \left\{ \left(\frac{1}{2}q_{xx}^* + F(|q|^2)q^* \right) R + \left(\frac{1}{2}q_{xx} + F(|q|^2)q \right) R^* \right\} dx \tag{2.53}$$

Equations (2.51)–(2.53) are known as the *modified integrals of motion*. In (2.51)–(2.53), setting $\epsilon = 0$ on the right side–the integrals of motion recovers. Now, using (2.51), (2.52), and (2.42) one can derive that

$$\frac{d\kappa}{dt} = \frac{\epsilon}{I_{0,2,0,0}} \frac{B}{A^2} \left[i \int_{-\infty}^{\infty} (q_x^*R - q_xR^*)dx - \kappa \int_{-\infty}^{\infty} (q^*R + qR^*)dx \right] \tag{2.54}$$

Again, differentiating (2.45) with respect to t and using (2.49) yields

$$v = \frac{d\bar{x}}{dt} = -\kappa + \frac{\epsilon}{I_{0,2,0,0}} \frac{B}{A^2} \int_{-\infty}^{\infty} x(q^*R + qR^*)dx \tag{2.55}$$

Finally, differentiating (2.49) with respect to t and subtracting from its conjugate, the following equation for the soliton wave number is obtained in the presence of perturbation terms

$$\frac{\partial\phi}{\partial t} = \frac{B^2}{2} \frac{I_{0,0,2,0}}{I_{0,2,0,0}} - \frac{\kappa^2}{2} + \frac{1}{I_{0,2,0,0}} \int_{-\infty}^{\infty} g^2(s)F(A^2g^2(s))ds$$
$$+ \frac{i\epsilon}{I_{0,2,0,0}} \frac{B}{2A^2} \int_{-\infty}^{\infty} (qR^* - q^*R)dx \tag{2.56}$$

Now, equations (2.51), (2.52), and (2.54)–(2.56) can also be rewritten in the following alternative forms

$$\frac{dE}{dt} = 2\epsilon \int_{-\infty}^{\infty} g(\tau)\Re[Re^{-i\phi}]d\tau \tag{2.57}$$

$$\frac{dM}{dt} = -\epsilon \int_{-\infty}^{\infty} \kappa g(\tau)\Re[Re^{-i\phi}] + 2A\frac{dg}{d\tau}\Im[Re^{-i\phi}]d\tau \tag{2.58}$$

$$\frac{d\kappa}{dt} = -\frac{2\epsilon}{I_{0,2,0,0}} \frac{\kappa}{A} \int_{-\infty}^{\infty} g(\tau)\Re[Re^{-i\phi}]d\tau$$

$$-\frac{2\epsilon}{I_{0,2,0,0}} \frac{1}{A} \int_{-\infty}^{\infty} \{\kappa g(\tau)\Re[Re^{-i\phi}] - B\frac{dg}{d\tau}\Im[Re^{-i\phi}]\}d\tau \qquad (2.59)$$

$$\frac{d\bar{x}}{dt} = -\kappa + \frac{2\epsilon}{I_{0,2,0,0}} \frac{1}{A} \int_{-\infty}^{\infty} x g(\tau)\Re[Re^{-i\phi}]d\tau \qquad (2.60)$$

$$\frac{\partial\phi}{\partial t} = \frac{B^2}{2} \frac{I_{0,0,2,0}}{I_{0,2,0,0}} - \frac{\kappa^2}{2} - \frac{1}{I_{0,2,0,0}} \int_{-\infty}^{\infty} g^2(s) F(A^2 g^2(s)) \, ds$$

$$+ \frac{\epsilon}{I_{0,2,0,0}} \frac{1}{A} \int_{-\infty}^{\infty} g(\tau)\Im[Re^{-i\phi}]d\tau \qquad (2.61)$$

where, once again, $\tau = B(t)(x - \bar{x})$, while $\phi = -\kappa x + \omega t + \sigma_0$. These alternative forms of the adiabatic evolution equations for the soliton parameters are known as the *amplitude-phase format*. Equations (2.57)–(2.61) give the adiabatic dynamics of the soliton parameters in the presence of perturbation terms.

In this book, a particular form of the perturbation term R will be studied, for various kinds of nonlinearities, that is given by [86]

$$R = \delta|q|^{2m}q + \sigma q \int_{-\infty}^{x} |q|^2 ds \qquad (2.62)$$

In (2.62), δ is the coefficient of nonlinear damping or amplification, depending on its sign, and m could be 0, 1, or 2. For $m = 0$, δ is the linear amplification or *attenuation* according to δ being positive or negative. For $m = 1$, δ represents the two-photon absorption (or a nonlinear gain if $\delta > 0$). If $m = 2$, δ gives a higher-order correction (saturation or loss) to the nonlinear amplification-absorption. Also, σ is the coefficient of saturable amplifiers that is introduced to compensate for the losses in optical fibers. A model with a saturation term included is more satisfactory from a physical point of view, since then the stable soliton propagation is ensured, in principle, over an indefinite propagation distance, including transoceanic distances.

Several factors contribute to nonlinear damping or attenuation. One is the material absorption. Rayleigh scattering is a fundamental loss mechanism that rises from random density fluctuations frozen into the fused silica in the manufacturing process and is a dominating factor. Pure silica absorbs either in the ultraviolet region or in the far-infrared region beyond 2 μm. However, even a relatively small amount of impurities can lead to significant absorption in the wavelength window 0.5–2 μm. One of the most important impurity factors affecting fiber loss is the hydroxide (OH) ion. Special precautions are taken during the fiber fabrication process to ensure that the OH impurity level is less than one part per 100 million.

So, the perturbed NLSE that is going to be considered is given by

$$iq_t + \frac{1}{2}q_{xx} + F(|q|^2)q = i\epsilon \left(\delta|q|^{2m}q + \sigma q \int_{-\infty}^{x} |q|^2 ds \right) \tag{2.63}$$

To obtain the parameter dynamics of the solitons caused by the particular form of perturbations given in (2.63), proceed as follows. Substitute R given by (2.63) into (2.51)–(2.56) and carry out the integration using the notation that was introduced in (2.40). The following equations are obtained [86]

$$\frac{dE}{dt} = \frac{2\epsilon}{B^2} \left[\delta A^{2m+2} B I_{0,2m+2,0,0} + \sigma \int_{-\infty}^{\infty} g^2(\tau) \left(\int_{-\infty}^{\tau} g^2(s)ds \right) d\tau \right] \tag{2.64}$$

$$\frac{dM}{dt} = \frac{2\epsilon\kappa}{B^2} \left[\delta A^{2m+2} B I_{0,2m+2,0,0} + \sigma \int_{-\infty}^{\infty} g^2(\tau) \left(\int_{-\infty}^{\tau} g^2(s)ds \right) d\tau \right] \tag{2.65}$$

$$\frac{d\kappa}{dt} = 0 \tag{2.66}$$

$$v = \frac{d\bar{x}}{dt} = -\kappa + \frac{2\epsilon\sigma}{I_{0,2,0,0}} \frac{A^2}{B^2} \int_{-\infty}^{\infty} \tau g^2(\tau) \left(\int_{-\infty}^{\tau} g^2(s)ds \right) d\tau \tag{2.67}$$

$$\frac{\partial\phi}{\partial t} = \frac{B^2}{2} \frac{I_{0,0,2,0}}{I_{0,2,0,0}} - \frac{\kappa^2}{2} + \frac{1}{I_{0,2,0,0}} \int_{-\infty}^{\infty} g^2(s) F(A^2 g^2(s))ds \tag{2.68}$$

The integrals in (2.64)–(2.68) will be explicitly calculated in the upcoming chapters for various laws of nonlinearity where the functions $F(\zeta)$ and $g(s)$ are exactly known. This will give the laws of adiabatic dynamics of the solitons in the presence of perturbation terms for various laws of nonlinearity.

2.5   Quasi-Stationary Solution

The idea of quasi-stationarity for solving the nonlinear evolution equations was first introduced in 1981 by Kodama and Ablowitz. Later, this idea was extended to study the NLSE, for Kerr law nonlinearity, with Hamiltonian and non-Hamiltonian type perturbations. Subsequently, it was also used to study the perturbed NLSE with various kinds of non-Kerr law nonlinearities. Studying the dynamics of solitons in the presence of perturbation terms gives more information than was obtained in the previous section by virtue of the soliton perturbation theory. For example, in order to integrate the perturbed NLSE up to $O(\epsilon)$, one needs to study the perturbed solitons by means of quasi-stationarity since the soliton perturbation theory fails to integrate the perturbed NLSE. The basic idea of a quasi-stationary (QS) method is explained in the following subsection [5, 228].

2.5.1 Mathematical Theory*

In a general setting, the solution of a perturbed nonlinear dispersive wave equation that is studied is of the type

$$K(q, q_t, q_x, \cdots) = \epsilon F(q, q_x, \cdots) \tag{2.69}$$

Here, K and F are nonlinear functions of q, q_x, $\cdots$, while $0 < \epsilon \ll 1$. The unperturbed equation (for $\epsilon = 0$)

$$K(q^{(0)}, q_t^{(0)}, q_x^{(0)}, \cdots) = 0 \tag{2.70}$$

has a solution $q^{(0)}$ that is taken as a solitary wave or a soliton solution. This solution in terms of certain natural fast and slow variables is written as

$$q^{(0)} = \hat{q}^{(0)}(\theta_1, \theta_2, \cdots \theta_m, T, X : P_1, P_2, \cdots, P_N) \tag{2.71}$$

where, θ_i for $1 \leq i \leq m$ are the so-called *fast variables* while $T = \epsilon t$ and $X = \epsilon x$ are the *slow variables*, and P_l for $1 \leq l \leq N$ is the parameter that depends on the slow variables. In many problems, only one fast variable, namely $\theta = x - P_1 t$ in the unperturbed problem, is needed. One can generalize θ to satisfy $\partial\theta/\partial x = 1$ and $\partial\theta/\partial t = -P_1$ and can use $P_1 = P_1(X, T)$ to remove the secular terms. With this, such a solution, given by (2.71), is a quasi-stationary solution and one can write $q = \hat{q}(\theta, X, T, \epsilon)$. It is necessary to develop equations for the parameters $P_1, \cdots, P_N$ by using appropiate conditions, such as the secularity conditions. There must be N such conditions. Some of these conditions are formed from Green's identity, as follows. Assuming an expression for $\hat{q}$ of the form

$$\hat{q} = \hat{q}^{(0)} + \epsilon\hat{q}^{(1)} + \cdots \tag{2.72}$$

(after introducing the appropiate variables θ_i, X, T, etc.). Then (2.70) is the leading-order problem and, assuming that K has only first order in time derivatives

$$L(\partial_{\theta_i}, \hat{q}^{(0)})\hat{q}^{(1)} = F(\hat{q}^{(0)}) - \frac{\partial K}{\partial q_t} \cdot q_T = F \tag{2.73}$$

is the first-order equation. Here, $L(\partial_{\theta_i}, \hat{q}^{(0)})$, $u = 0$ is the linearized equation of $K(q, q_t, q_x, \cdots) = 0$ after (x, t) is transformed to the appropiate coordinate θ_i. Calling v_i the M solutions of the homogenous adjoint problem satisfying the necessary boundary conditions (e.g., $v_i \to 0$, as $|\theta_i| \to \infty$)

$$L^A v_i = 0 \tag{2.74}$$

---

* Portions of Section 2.5.1 are reprinted with permission from Ablowitz, M.J. & H. Segur, *Solitons and the Inverse Scattering Transform*. 1981. Copyright© Society for Industrial and Applied Mathematics.

for $1 \leq i \leq M$ and $M \leq N$ with L^A being the adjoint operator to L, one forms

$$L\hat{q}^{(1)} \cdot v_i - (L^A v_i) \cdot \hat{q}^{(1)} = \hat{F} v_i \tag{2.75}$$

The left side of (2.73) is always a divergence (Green's theorem). It may be integrated to give the secularity conditions, also known as the Fredholm's alternative (FA). These secularity conditions enable us to compute a solution $\hat{q}^{(1)}$ to (2.73) that satisfies suitable boundary conditions (e.g., $\hat{q}^{(1)}$ is bounded as $|\theta| \to \infty$). However, as is standard in perturbation problems, there is still freedom in the solution. This is due to the fact that some terms in the solution $\hat{q}^{(1)}$ can be absorbed in the leading-order solution $\hat{q}^{(0)}$ by shifting other parameters. The solution $\hat{q}^{(1)}$ can be made unique by imposing additional conditions that reflect specific initial conditions or other normalizations. Continuation to higher order $\hat{q}^{(N)}$ is straightforward.

2.5.2   Application

In this book, a quasi-stationary solution to (2.63) will be obtained using the method that was discussed in the previous subsection. The main part of this work is to implement a perturbation scheme to solve (2.63) as follows [86]:

$$q = \hat{q}(\theta, T, X; \epsilon) e^{\frac{i}{\epsilon} \rho(T, X; \epsilon)} \tag{2.76}$$

where

$$\frac{\partial \theta}{\partial x} = 1, \quad \frac{\partial \theta}{\partial t} = 0$$

and

$$T = \epsilon t \quad X = \epsilon x$$

Here, as mentioned, θ is a fast variable while X and T are slow variables in space and time, respectively. When the perturbation terms of the NLSE are turned on, the soliton parameters A and B are then slowly varying functions, namely $A = A(X, T)$ and $B = B(X, T)$.

Substituting (2.76) into (2.63) and expanding

$$\hat{q} = \hat{q}^{(0)} + \epsilon \hat{q}^{(1)} + \epsilon^2 \hat{q}^{(2)} + \cdots$$

$$\rho = \rho^{(0)} + \epsilon \rho^{(1)} + \epsilon^2 \rho^{(2)} + \cdots$$

$$v = v^{(0)} + \epsilon v^{(1)} + \epsilon^2 v^{(2)} \cdots$$

gives, at the leading order

$$-\left\{ \rho_T^{(0)} + \frac{1}{2} (\rho_X^{(0)})^2 \right\} \hat{q}^{(0)} + \frac{1}{2} \frac{\partial^2 \hat{q}^{(0)}}{\partial \theta^2} + \hat{q}^{(0)} F[(\hat{q}^{(0)})^2] = 0 \tag{2.77}$$

and

$$\left(\rho_X^{(0)} - v^{(0)}\right)\frac{\partial \hat{q}^{(0)}}{\partial \theta} = 0 \tag{2.78}$$

from the real and imaginary parts. Now, (2.78) implies

$$\rho_X^{(0)} = v^{(0)} \tag{2.79}$$

Setting

$$h(B^2) = \rho_T^{(0)} + \frac{1}{2}\left(\rho_X^{(0)}\right)^2 = \rho_T^{(0)} + \frac{1}{2}(v^{(0)})^2 \tag{2.80}$$

where the function h depends on the nonlinearity F. Thus, (2.77) changes to

$$-h(B^2)\hat{q}^{(0)} + \frac{1}{2}\frac{\partial^2 \hat{q}^{(0)}}{\partial \theta^2} + \hat{q}^{(0)}F[(\hat{q}^{(0)})^2] = 0 \tag{2.81}$$

whose solution is (on comparing with [2.35])

$$\hat{q}^{(0)} = Ag[B(\theta - \bar{\theta})] \tag{2.82}$$

where

$$\frac{d\bar{\theta}}{dt} = v \tag{2.83}$$

At $O(\epsilon)$, decompose $\hat{q}^{(1)} = \hat{\phi}^{(1)} + i\hat{\psi}^{(1)}$ into its real and imaginary parts. Now, the equations for $\hat{\phi}^{(1)}$ and $\hat{\psi}^{(1)}$, by virtue of (2.77), are respectively

$$-h(B^2)\hat{\phi}^{(1)} + \frac{1}{2}\frac{\partial^2 \hat{\phi}^{(1)}}{\partial \theta^2} + 2(\hat{q}^{(0)})^2\hat{\phi}^{(1)}F'[(\hat{q}^{(0)})^2] + \hat{\phi}^{(1)}F[(\hat{q}^{(0)})^2]$$
$$= \{\rho_T^{(1)} + v^{(0)}\rho_X^{(1)}\}\hat{q}^{(0)} - \frac{\partial^2 \hat{q}^{(0)}}{\partial \theta \partial X} \tag{2.84}$$

and

$$-h(B^2)\hat{\psi}^{(1)} + \frac{1}{2}\frac{\partial^2 \hat{\psi}^{(1)}}{\partial \theta^2} + F[(\hat{q}^{(0)})^2]\hat{\psi}^{(1)}$$
$$= -\frac{\partial \hat{q}^{(0)}}{\partial T} - v^{(0)}\frac{\partial \hat{q}^{(0)}}{\partial X} - \{\rho_X^{(1)} - v^{(1)} + \sigma\}\frac{\partial \hat{q}^{(0)}}{\partial \theta}$$
$$- \rho_{XX}^{(0)}\hat{q}^{(0)} - \delta(\hat{q}^{(0)})^{2m+1} + \sigma\hat{q}^{(0)}\int_{-\infty}^{x}(\hat{q}^{(0)})^2ds \tag{2.85}$$

Equations (2.84) and (2.85) contain secular terms that are also known as resonances. In order to eliminate these secular terms, to avoid unbounded growth, one needs to use the FA, which states that the right side of the ODEs in (2.84) and (2.85) should be orthogonal to the null space of the adjoint operator of the left side. But for the ODEs given by (2.84) and (2.85), the left side is self-adjoint.

Thus, FA applied to (2.84) yields

$$\int_{-\infty}^{\infty} \left[\{\rho_T^{(1)} + v^{(0)} \rho_X^{(1)}\} \hat{q}^{(0)} - \frac{\partial^2 \hat{q}^{(0)}}{\partial \theta \partial X} \right] \frac{\partial \hat{q}^{(0)}}{\partial \theta} d\theta = 0 \qquad (2.86)$$

which leads to

$$B \frac{\partial A}{\partial X} I_{0,0,2,0} + A \frac{\partial B}{\partial X} I_{0,0,2,0} + A \frac{\partial B}{\partial X} I_{1,0,0,1} = 0 \qquad (2.87)$$

Similarly, with the second homogenous solution of (2.73), one recovers

$$\rho_T^{(1)} + v^{(0)} \rho_X^{(1)} = 0 \qquad (2.88)$$

whereas, FA applied to (2.85) gives

$$\int_{-\infty}^{\infty} \left[\frac{\partial \hat{q}^{(0)}}{\partial T} + v^{(0)} \frac{\partial \hat{q}^{(0)}}{\partial X} + \{\rho_X^{(1)} - v^{(1)} + \sigma\} \frac{\partial \hat{q}^{(0)}}{\partial \theta} \right.$$
$$\left. + \rho_{XX}^{(0)} \hat{q}^{(0)} + \delta(\hat{q}^{(0)})^{2m+1} - \sigma \hat{q}^{(0)} \int_{-\infty}^{x} (\hat{q}^{(0)})^2 ds \right] \hat{q}^{(0)} d\theta = 0 \qquad (2.89)$$

which yields

$$B \frac{\partial A}{\partial T} I_{0,2,0,0} + A \frac{\partial B}{\partial T} I_{1,1,1,0} + v^{(0)} B \frac{\partial A}{\partial X} I_{0,0,2,0} + v^{(0)} A \frac{\partial B}{\partial X} I_{1,1,1,0}$$
$$= \delta A^{2m+1} B I_{0,2m+2,0,0} - \rho_{XX}^{(0)} AB I_{0,0,2,0} + \sigma \hat{q}^{(0)} \int_{-\infty}^{x} (\hat{q}^{(0)})^2 ds \qquad (2.90)$$

Similarly, with the second homogenous solution of (2.85), one gets

$$\rho_X^{(1)} = v^{(1)} - \sigma \qquad (2.91)$$

Since $A(t)$ and $B(t)$ are related, depending on the functional form of $F(s)$, (2.87) leads to the conclusion

$$\frac{\partial A}{\partial X} = \frac{\partial B}{\partial X} = 0 \qquad (2.92)$$

so that A and B are functions of T only.

Also, in an ideal soliton-based, fiber-optic communication system, input pulses launched into the fiber should be unchirped in order to avoid shedding part of the pulse energy as a dispersive tail during the process of soliton formation [22]. So, in (2.90), using (2.92) and setting $\rho_{XX}^{(0)} = 0$ to eliminate frequency chirp gives

$$B \frac{dA}{dT} I_{0,2,0,0} + A \frac{dB}{dT} I_{1,1,1,0} = \delta A^{2m+1} B I_{0,2m+2,0,0} + \sigma \hat{q}^{(0)} \int_{-\infty}^{x} (\hat{q}^{(0)})^2 ds$$
$$(2.93)$$

It is to be noted here that (2.93) can be recovered by the usual soliton perturbation theory (SPT), provided the type of nonlinearity in F is known. However, the relations (2.87), (2.88), and (2.91) cannot be obtained by the SPT. This is where the SPT fails. Thus, a more general approach to study optical soliton perturbation is the QS method. Now (2.84), by virtue of (2.87) and (2.90), reduces to

$$-h(B^2)\hat{\phi}^{(1)} + \frac{1}{2}\frac{\partial^2\hat{\phi}^{(1)}}{\partial\theta^2} + 2(\hat{q}^{(0)})^2\hat{\phi}^{(1)}F'[(\hat{q}^{(0)})^2] + \hat{\phi}^{(1)}F[(\hat{q}^{(0)})^2] = 0 \qquad (2.94)$$

while (2.85), by virtue of (2.87) and (2.91), gives

$$-h(B^2)\hat{\psi}^{(1)} + \frac{1}{2}\frac{\partial^2\hat{\psi}^{(1)}}{\partial\theta^2} + F[(\hat{q}^{(0)})^2]\hat{\psi}^{(1)}$$
$$= -\frac{\partial\hat{q}^{(0)}}{\partial T} + \delta(\hat{q}^{(0)})^{2m+1} + \sigma\hat{q}^{(0)}\int_{-\infty}^{x}(\hat{q}^{(0)})^2 ds \qquad (2.95)$$

The solutions to (2.94) and (2.95) are respectively

$$\hat{\phi}^{(1)} = 0 \qquad (2.96)$$

and

$$\begin{aligned}
\hat{\psi}^{(1)} = {} & \frac{2A}{B}\frac{\partial\bar{\theta}}{\partial T}g(\tau)\int^{\tau}\frac{1}{g^2(s_2)}\left(\int^{s_2}g(s_1)g'(s_1)ds_1\right)ds_2 \\
& -\frac{2}{B^2}\frac{dA}{dT}g(\tau)\int^{\tau}\frac{1}{g^2(s_2)}\left(\int^{s_2}g^2(s_1)ds_1\right)ds_2 \\
& -\frac{2A}{B^3}\frac{dB}{dT}g(\tau)\int^{\tau}\frac{1}{g^2(s_2)}\left(\int^{s_2}s_1 g(s_1)g'(s_1)ds_1\right)ds_2 \\
& +2\delta\frac{A^{2m+1}}{B^2}g(\tau)\int^{\tau}\frac{1}{g^2(s_2)}\left(\int^{s_2}g^{2m+2}(s_1)ds_1\right)ds_2 \\
& +2\sigma\frac{A^3}{B^2}g(\tau)\int^{\tau}\frac{1}{g^2(s_3)}\left(\int^{s_3}g^2(s_2)\left(\int^{s_2}g^2(s_1)ds_1\right)ds_2\right)ds_3 \qquad (2.97)
\end{aligned}$$

The QS solution of (2.63) finally is

$$q \approx Pe^{i\psi} \qquad (2.98)$$

where

$$P = \hat{q}^{(0)} \qquad (2.99)$$

and

$$\psi = \epsilon Q(\theta) + \frac{1}{\epsilon}\rho(X, T) \qquad (2.100)$$

with

$$Q(\theta) = \hat{\psi}^{(1)}/\hat{q}^{(0)} \tag{2.101}$$

Equation (2.98) is, thus, the "formal" solution to the perturbed NLSE that is given by (2.63). Note that this solution is a general one in the sense that the law of nonlinearity given by $F(s)$ is not yet known. However, in the following chapters, the special cases of this law of nonlinearity will be considered. Once the functional form of $F(s)$ is known, it will be possible to evaluate the integrals seen in (2.97).

Exercises

1. Prove that the Hamiltonian given by (2.31) is a conserved quantity.
2. Derive the wave number for the soliton that is given by (2.39).
3. Use the laws of adiabatic deformation of energy and linear momentum of the soliton that are given by (2.51) and (2.52), respectively, to derive the adiabatic dynamics of the soliton frequency that is given by (2.54).
4. Use the expression for the velocity of the soliton that is given by (2.55) to derive the expression in (2.67) corresponding to the perturbation terms given by (2.62).
5. Use the adiabatic dynamics for the soliton wave number given by (2.56) to derive the expression in (2.68) corresponding to the perturbation terms that are given in (2.62).

3

Kerr Law Nonlinearity

This chapter talks about the detailed aspects of optical solitons that are governed by the nonlinear Schrödinger's equation (NLSE) with Kerr law nonlinearity, which is also commonly known as the *cubic Schrödinger's equation*. Section 3.1 talks about the physics of the origin of Kerr law nonlinearity. Section 3.2 contains an introductory discussion about the technique of inverse scattering transform (IST) that is used to integrate the NLSE with Kerr law nonlinearity. Moreover, the comparison between the IST and the Fourier transform (FT) is discussed here. Also, the 1-soliton solution is given. The infinitely many conserved quantities of the NLSE are talked about in Section 3.3 along with the Hamiltonian structure, while Section 3.4 leads to a discussion about another technique of solving the NLSE, namely the variational principle using the Lagrangian. Section 3.5 is about the quasi-stationarity applied to the case of Kerr law nonlinearity, and Section 3.6 introduces the Lie transform technique that can be used to integrate the perturbed NLSE, for Kerr law, that contains Hamiltonian perturbation terms only.

3.1 Introduction

In this chapter, equation (2.1) will be studied for a special case, namely when $F(s) = s$, which is known as the *Kerr law of nonlinearity*. In this case, (2.1) is integrable by the method of the IST that was first applied to solve (2.1) by Zakharov and Shabat [296] in 1972. The IST is the nonlinear analog of the Fourier transform that is used for solving linear partial differential equations. The IST will be discussed in the next section of this chapter.

The Kerr law of nonlinearity originates from the fact that a light wave in an optical fiber faces nonlinear responses from nonharmonic motion of electrons bound in molecules, caused by an external electric field. Even though the nonlinear responses are extremely weak, their effects appear in various ways over long distances of propagation that are measured in terms of light wavelength.

The origin of nonlinear response is related to the nonharmonic motion of bound electrons under the influence of an applied field. As a result, the induced polarization (**P**) is not linear in the electric field (**E**) but involves higher order terms in the electric field amplitude as [185, 192, 311]

$$\begin{aligned} \mathbf{P} = \epsilon_0 \Bigg[& \int_{-\infty}^{\infty} \chi^{(1)}(t - \tau)\mathbf{E}(\tau)d\tau \\ & + \int_{-\infty}^{\infty}\int_{-\infty}^{\infty} \chi^{(2)}(t - \tau_1)\chi^{(2)}(t - \tau_2)\mathbf{E}(\tau_1)\mathbf{E}(\tau_2)d\tau_1 d\tau_2 \\ & + \int_{-\infty}^{\infty}\int_{-\infty}^{\infty}\int_{-\infty}^{\infty} \chi^{(3)}(t - \tau_1)\chi^{(3)}(t - \tau_2)\chi^{(3)}(t - \tau_3)\mathbf{E}(\tau_1)\mathbf{E}(\tau_2)\mathbf{E}(\tau_3) \\ & d\tau_1 d\tau_2 d\tau_3 + \cdots \Bigg] \end{aligned}$$
(3.1)

where ϵ_0 is the permittivity of the vacuum and $\chi^{(j)}$, for $j = 1, 2, \ldots$ is the jth order susceptibility tensor of rank $j + 1$, which accounts for light polarization effects. The linear susceptibility, $\chi^{(1)}$, gives the linear refractive index n_0 and the attenuation coefficient. The second order susceptibility, $\chi^{(2)}$, is responsible for the second-harmonic generation. However, it is generally negligible for silicon dioxide, which has an inversion symmetry at the molecular level, unless $\chi^{(2)}$ contains some resonant effects. As a result, optical fibers do not normally exhibit second-order nonlinear effects, although the electric-quadrupole and magnetic-dipole moments can generate weak second-order nonlinear effects. The lowest nonlinear effects in nonlinear fibers originate primarily from the third order susceptibility, $\chi^{(3)}$. A light wave with frequency ω sees nonlinear response of the $\chi^{(3)}$ term through the interaction of ω, $-\omega$, and ω components. The response contributes to the nonlinear modification of the index of refraction [22, 192, 227]

$$n(\omega, |E|^2) = n_0(\omega) + n_2(\omega)|E|^2$$
(3.2)

where E is the amplitude of the wave electric field and n_2 is related to $\chi^{(3)}$ through

$$n_2(\omega) = \frac{3}{8n_0}\chi^{(3)}_{xxxx}$$
(3.3)

for a linearly polarized wave in the x direction. Here n_0 is the linear refractive index of the medium and the nonlinear index of refraction n_2 is called the *Kerr coefficient* and has a value of approximately 10^{-22} m^2/W for silica fibers. Even though the fiber nonlinearity is small, the nonlinear effects accumulate over long distances and can have a significant impact due to the high intensity of the light wave over a small fiber cross-section. Since the Kerr effect here originates from the nonharmonic motion of electrons bound in molecules, the instantaneous response (3.3) is satisfied. In a fiber, the electric field magnitude varies in its cross-section, thus a proper averaging of $|E|^2$ should be taken into account when evaluating the response. By itself, the Kerr nonlinearity produces an intensity-dependent phase shift that results in spectral broadening during propagation. For the Kerr law of nonlinearity in equation (2.1), $F(s) = s$ and so $f(s) = s^2/2$. In the following subsection, the derivation of the NLSE with Kerr law of nonlinearity is given.

3.1.1 The Nonlinear Schrödinger's Equation

In this subsection, the derivation of the dynamical equation of an optical pulse propagating through an optical fiber is given for the case of the Kerr law of nonlinearity. The electric field $\vec{E}(\vec{r}, t)$ of the pulse is governed by the wave equation

$$\vec{\nabla} \times \vec{\nabla} \times \vec{E}(\vec{r}, t) + \mu_0 \frac{\partial^2 \vec{P}(\vec{r}, t)}{\partial t^2} + \frac{1}{c^2} \frac{\partial^2 \vec{E}(r, t)}{\partial t^2} = 0 \qquad (3.4)$$

where $\vec{P}(\vec{r}, t)$ is the induced polarization in the media, μ_0 is the vacuum permeability, and c is the velocity of light. The induced polarization consists of two parts: the linear part $\vec{P}_L (= \epsilon_0 \chi^{(1)} \vec{E})$ and the nonlinear part P_{nL}; ϵ_0 is the vacuum permittivity and $\chi^{(1)}$ is the linear susceptibility. The optical pulse is taken to be of the form

$$\vec{E}(\vec{r}, t) = \frac{1}{2} \vec{x} [\vec{E}(\vec{r}, t) e^{i\omega_0 t} + cc] \qquad (3.5)$$

$$\vec{P}_L(\vec{r}, t) = \frac{1}{2} \vec{x} [\vec{P}_L(\vec{r}, t) e^{i\omega_0 t} + cc] \qquad (3.6)$$

$$\vec{P}_{nL}(\vec{r}, t) = \frac{1}{2} \vec{x} [\vec{P}_{nL}(\vec{r}, t) e^{i\omega_0 t} + cc] \qquad (3.7)$$

where ω_0 is the carrier frequency of the light wave, $\vec{x}$ is the polarization unit vector, and $\vec{E}(\vec{r}, t)$, $\vec{P}_L(\vec{r}, t)$, and $\vec{P}_{nL}(\vec{r}, t)$ are slowly varying functions of time. In (3.5)–(3.7), cc represents the complex conjugate. For optical fibers, the first term of (3.4) can be approximated as

$$\vec{\nabla} \times \vec{\nabla} \times \vec{E}(\vec{r}, t) = \vec{\nabla}(\vec{\nabla} \cdot \vec{E}(\vec{r}, t)) - \nabla^2 \vec{E}(\vec{r}, t) \approx -\nabla^2 \vec{E}(\vec{r}, t) \qquad (3.8)$$

so that (3.4) reduces to

$$\nabla^2 \vec{E}(\vec{r}, t) - \mu_0 \frac{\partial^2 \vec{P}(\vec{r}, t)}{\partial t^2} - \frac{1}{c^2} \frac{\partial^2 \vec{E}(r, t)}{\partial t^2} = 0 \qquad (3.9)$$

At this stage, it is necessary to make one more simplification. It is assumed that the linear part of the polarization depends on the frequency, while the nonlinear part is independent of the frequency. In other words, only those media for which the response of nonlinearity is instantaneous are considered. The nonlinear polarization for optical fibers, which are oscillating at ω_0, is of the form

$$\vec{P}_{nL}(\vec{r}, t) = \frac{3}{4} \epsilon_0 \chi^{(3)} |E|^2 \vec{E}(\vec{r}, t) \qquad (3.10)$$

The Fourier transform of $\vec{E}(\vec{r}, t)$ is now defined as

$$\tilde{E}(\vec{r}, \omega) = \int_{-\infty}^{\infty} \vec{E}(\vec{r}, t) e^{i(\omega - \omega_0)t} dt \tag{3.11}$$

Treating the nonlinear part of susceptibility as a constant, the Fourier transform of (3.9) gives

$$\nabla^2 \tilde{E}(r, \omega - \omega_0)$$
$$+ \frac{\omega^2}{c^2} \{1 + \chi^{(1)}(\omega - \omega_0)\} \tilde{E}(r, \omega - \omega_0)$$
$$+ \frac{\omega^2}{c^2} \left(\frac{3}{4} \chi^{(3)} |E|^2 \right) \tilde{E}(r, \omega - \omega_0) = 0 \tag{3.12}$$

The frequency-dependent wave number k is defined through the relation

$$k^2(\omega) = \frac{\omega^2}{c^2} \{1 + \chi^{(1)}(\omega - \omega_0)\} \tag{3.13}$$

From (3.13), expanding k^2 in a Taylor series about the carrier frequency ω_0 gives

$$k^2 = k_0^2 + 2(\omega - \omega_0) k_0 k_0' + (\omega - \omega_0)^2 (k_0')^2 + (\omega - \omega_0)^2 k_0 k_0'' \tag{3.14}$$

where the prime in (3.14) denotes a derivative with respect to ω and evaluated at ω_0 and

$$k_0 = \frac{\omega_0}{c} \sqrt{1 + \chi^{(1)}(\omega_0)} \tag{3.15}$$

Also k_0' is the inverse of group velocity v_g, while k_0'' is the group velocity dispersion. For a normal dispersion regime $k_0'' > 0$, while in the anomalous dispersion regime $k_0'' < 0$. Inserting (3.14) into (3.12) and then transforming the equation in the time domain gives

$$\nabla^2 E + k_0^2 E + 2i k_0 k_0' \frac{\partial E}{\partial t} - (k_0')^2 \frac{\partial^2 E}{\partial t^2} - k_0 k_0'' \frac{\partial^2 E}{\partial t^2} + \frac{3}{4} \chi^{(3)} \frac{\omega_0^2}{c^2} |E|^2 E = 0 \tag{3.16}$$

Choosing the ansatz $E(\vec{r}, t) = A(\vec{r}, t) e^{ik_0 z}$, where the scalar envelope $A(r, t)$ of the optical pulse varies slowly over one optical period, one can introduce the slowly varying envelope approximation as

$$\frac{\partial^2 A}{\partial t^2} < \omega_0 \frac{\partial A}{\partial t} \tag{3.17}$$

and

$$\frac{\partial^2 A}{\partial z^2} < k_0 \frac{\partial A}{\partial z} \tag{3.18}$$

Substituting the ansatz for $E(r, t)$ and employing the slowly varying envelope approximation gives

$$\nabla_\perp^2 A + 2ik_0 \frac{\partial A}{\partial z}$$

$$+ \frac{\partial^2 A}{\partial z^2} + 2ik_0 k_0' \frac{\partial A}{\partial t} - k_0 k_0'' \frac{\partial^2 A}{\partial t^2} - (k_0')^2 \frac{\partial^2 A}{\partial t^2} + \frac{3}{4} \chi^{(3)} \frac{\omega_0^2}{c^2} |A|^2 A = 0 \qquad (3.19)$$

where $\nabla_\perp^2$ is the perpendicular Laplacian. Now, changing to a frame that is moving with a group velocity v_g of the pulse and introducing the transformations

$$z' = z \qquad (3.20)$$

and

$$t' = t - \frac{z}{v_g} = t - zk_0' \qquad (3.21)$$

gives

$$\nabla_\perp^2 A + \frac{\partial^2 A}{\partial z^2} + 2ik_0 \frac{\partial A}{\partial z} - k_0 k_0'' \frac{\partial^2 A}{\partial t^2} + \frac{3}{4} \chi^{(3)} \frac{\omega_0^2}{c^2} |A|^2 = 0 \qquad (3.22)$$

In writing the above equation, the primes over the variables are dropped simply for convenience. The second term is much smaller due to the slowly varying envelope. Thus (3.22) can be modified to

$$\nabla_\perp^2 A + 2ik_0 \frac{\partial A}{\partial z} - k_0 k_0'' \frac{\partial^2 A}{\partial t^2} + \frac{3}{4} \chi^{(3)} \frac{\omega_0^2}{c^2} |A|^2 A = 0 \qquad (3.23)$$

Now consider only the linear part of equation (3.23), neglecting the fourth term. It is assumed that

$$A(r, t) = R(r) Q(z, t) e^{-i\delta z} \qquad (3.24)$$

where $R(r)$ is the transverse mode profile and δ is the corresponding eigenvalue satisfying

$$\nabla_\perp^2 R(r) + 2\delta k_0 R = 0 \qquad (3.25)$$

For the nonlinear equation (3.23), it is assumed that (3.25) still holds for the transverse mode structure of the propagating optical pulse. Therefore, inserting (3.24) in (3.23) and employing (3.25), one can finally arrive at

$$i\frac{\partial Q}{\partial z} - \frac{k_0''}{2} \frac{\partial^2 Q}{\partial t^2} + \frac{\omega}{c} n_2 |R|^2 |Q|^2 Q = 0 \qquad (3.26)$$

Averaging this equation over a fiber cross-section gives

$$i\frac{\partial Q}{\partial z} - \frac{k_0''}{2} \frac{\partial^2 Q}{\partial t^2} + \frac{\omega}{c} n_2 \alpha |Q|^2 Q = 0 \qquad (3.27)$$

where

$$\alpha = \frac{\int |R|^4 dr}{\int |R|^2 dr} \tag{3.28}$$

The numerical value of α depends on the concrete form of $R(r)$. For a Gaussian field profile in optical fibers, $\alpha = 0.5$. In subsequent discussions, only anomalous media are considered; $k_0'' < 0$ and so

$$\tau = \frac{t}{t_0} \tag{3.29}$$

$$Q = \sqrt{P_0} q \tag{3.30}$$

$$L_d = \frac{t_0^2}{|k_0''|} \tag{3.31}$$

and

$$P_0 = \frac{|k_0''|c}{\omega_0 n_2 \alpha t_0^2} \tag{3.32}$$

which transforms (3.27) to

$$i\frac{\partial q}{\partial z} + \frac{1}{2}\frac{\partial^2 q}{\partial \tau^2} + |q|^2 q = 0 \tag{3.33}$$

This equation is known as the NLSE and has been widely used for the propagation of solitons through optical fiber with Kerr law of nonlinearity. In order to maintain the uniformity of the notation in this book, the changes $z \to t$ and $\tau \to x$ are utilized so that the NLSE can be written in the standard notation as

$$iq_t + \frac{1}{2}q_{xx} + |q|^2 q = 0 \tag{3.34}$$

3.2 Traveling Wave Solution

For the case of Kerr law nonlinearity, equation (2.1) reduces to

$$iq_t + \frac{1}{2}q_{xx} + |q|^2 q = 0 \tag{3.35}$$

so that equation (2.10) simplifies to

$$B^2 g'' - (\kappa^2 - 2\omega)g + 2A^2 g^3 = 0 \tag{3.36}$$

Multiplying equation (3.36) by g' and integrating yields

$$(g')^2 = \frac{g^2(\kappa^2 - 2\omega - A^2 g^2)}{B^2} \tag{3.37}$$

Separating variables and integrating yields, on choosing the integration constant to be zero

$$x - vt = \int \frac{dg}{g\sqrt{\kappa^2 - 2\omega - A^2 g^2}} \tag{3.38}$$

On substituting

$$g^2 = \frac{\kappa^2 - 2\omega}{A^2 \cosh^2 \theta} \tag{3.39}$$

leads to the soliton solution

$$q(x, t) = \frac{A}{\cosh[B(x - \bar{x}(t))]} e^{i(-\kappa x + \omega t + \sigma_0)} \tag{3.40}$$

where

$$\kappa = -v \tag{3.41}$$

and

$$\omega = \frac{B^2 - \kappa^2}{2} \tag{3.42}$$

while

$$A = B \tag{3.43}$$

This is the 1-soliton solution of the NLSE with Kerr law nonlinearity that can be obtained by traveling wave ansatz.

3.3   Inverse Scattering Transform

In this section, the IST, a powerful method of integrating the NLSE with Kerr law nonlinearity that was used for solving (3.34), will be briefly presented. Subsequently, the 1-soliton solution of (3.34) will be derived. A detailed study of IST can be found in various books. Among the best books is the famous book by Ablowitz and Clarkson [6]. Basically, the IST is schematically similar to the FT that is used for solving linear differential equations. The FT can be described schematically as

$$q(x, 0) \xrightarrow{FT} \hat{q}(\omega, 0)$$

$$\text{time evolution} \downarrow \qquad\qquad \text{time evolution} \downarrow$$

$$q(x, t) \xleftarrow{\text{inverse } FT} \hat{q}(\omega, t)$$

while the IST, schematically, is

$$q(x, 0) \quad \xrightarrow{\ scattering\ } \quad \Sigma(x, 0)$$

$$time\ evolution \downarrow \qquad\qquad time\ evolution \downarrow$$

$$q(x, t) \quad \xleftarrow{\ IST\ } \quad \Sigma(x, t)$$

The main idea of the IST consists in recognizing that equation (3.34) can be expressed as the compatibility conditions of the linear equations for the wave function $\psi(x, t; \zeta)$ [384]

$$L(t)\psi = \zeta \psi \qquad\qquad (3.44)$$

and

$$\frac{\partial \psi}{\partial t} = M(t)\psi \qquad\qquad (3.45)$$

where L and M are the differential operators that formulate the Lax pair of the NLSE that are given by

$$L = \begin{bmatrix} i\frac{\partial}{\partial x} & q \\ -q^* & -i\frac{\partial}{\partial x} \end{bmatrix} \qquad\qquad (3.46)$$

and

$$M = \begin{bmatrix} -i\zeta^2 + \frac{i}{2}|q|^2 & -iq\zeta - \frac{1}{2}\frac{\partial q}{\partial x} \\ -iq^*\zeta + \frac{1}{2}\frac{\partial q^*}{\partial x} & i\zeta^2 - \frac{i}{2}|q|^2 \end{bmatrix} \qquad\qquad (3.47)$$

Equation (3.44) is an eigenvalue problem with eigenvalue ζ. This equation can also be rewritten as the x-evolution of the wave function, namely

$$\frac{\partial \psi}{\partial x} = P\psi \qquad\qquad (3.48)$$

where

$$P = \begin{bmatrix} -i\zeta & iq \\ iq^* & i\zeta \end{bmatrix} \qquad\qquad (3.49)$$

On the other hand, equations (3.45) determine the time evolution of the wave function ψ. Also, the eigenvalue ζ is *isospectral*, namely

$$\frac{d\zeta}{dt} = 0 \qquad\qquad (3.50)$$

For consistency between (3.45) and (3.48), the compatibility condition

$$\frac{\partial^2 \psi}{\partial x \partial t} = \frac{\partial^2 \psi}{\partial t \partial x} \tag{3.51}$$

is equivalent to

$$\frac{\partial P}{\partial t} - \frac{\partial M}{\partial x} + [P, M] = 0 \tag{3.52}$$

where in (3.52) the notations mean

$$[P, M] = PM - MP \tag{3.53}$$

is the commutator of P and M. This equation gives the NLSE with Kerr law nonlinearity. The advantage of this representation is that one can solve the nonlinear equation (3.34) by virtue of two linear problems. In fact, one associates the initial pulse profile $q(x, 0)$, the scattering data (similar to the Fourier coefficients) that are obtained, as follows, from the solution of equation (3.44). The wave functions are defined as

$$\Psi = \begin{pmatrix} \psi_1 \\ \psi_2 \end{pmatrix} \tag{3.54}$$

and

$$\Phi = \begin{pmatrix} \phi_1 \\ \phi_2 \end{pmatrix} \tag{3.55}$$

with the asymptotic values for $\Re(\zeta) = \xi$

$$\Phi(x;\xi) \to \begin{pmatrix} 1 \\ 0 \end{pmatrix} e^{-i\xi x} \quad as \quad x \to -\infty \tag{3.56}$$

$$\Psi(x;\xi) \to \begin{pmatrix} 0 \\ 1 \end{pmatrix} e^{i\xi x} \quad as \quad x \to \infty \tag{3.57}$$

where $\Re(\zeta)$ represents the real part of ζ. It can be shown that $\Phi(x;\xi)$ and $\Psi(x;\xi)$ can be analytically extended to the upper half plane of ξ. The pair of solutions $\Psi, \tilde{\Psi}$ forms a complete system of solutions to (3.44) where

$$\tilde{\Psi} = \begin{pmatrix} \psi_2^* \\ -\psi_1^* \end{pmatrix} \tag{3.58}$$

Thus, any other solution can be expressed as its linear combination, namely

$$\Phi(x;\xi) = a(\xi)\tilde{\Psi} + b(\xi)\Psi(x;\xi) \tag{3.59}$$

with the limit as $x \to \infty$, one obtains from this equation

$$\Phi(x;\xi) = a(\xi) \to \begin{pmatrix} 1 \\ 0 \end{pmatrix} e^{-i\xi x} + b(\xi) \to \begin{pmatrix} 0 \\ 1 \end{pmatrix} e^{i\xi x} \qquad (3.60)$$

From this couple of equations, one can see that $1/a(\xi)$ and $b(\xi)/a(\xi)$ respectively represent the transmission and reflection coefficient. Also, from equation (3.60), one can write

$$a(\xi) = W(\phi, \psi)(\xi) \qquad (3.61)$$

and

$$a(\xi) = W(\phi, \psi^*)(\xi) \qquad (3.62)$$

where the x-dependence of the Wronskian $W(f, g) = f_1 g_2 - f_2 g_1$ is used. It is to be noted that $a(\zeta)$ can be analytically extended to the upper half plane of ζ and also asymptotically $a(\zeta) \to 1$ as $|\zeta| \to \infty$.

The discrete eigenvalues of (3.44) corresponding to the points $\zeta = \zeta_n$ for $n = 1 \ldots N$; $\Im(\zeta) > 0$, where $a(\zeta) = 0$. Then the eigenfunctions obey

$$\Phi(x;\zeta_n) = b_n \Psi(x;\zeta_n) \qquad (3.63)$$

This equation shows that both Φ and Ψ approach zero as in the limit as $|x| \to \infty$. In other words, the discrete eigenvalues correspond to the bound states of the scattering problem (3.44) or solitons.

For any given potential function $q(x, 0)$, one can then compute the solution of the direct scattering problem (3.45) with the scattering data

$$\sum(t = 0) = [r(\xi, 0), \Im(\xi) = 0, \zeta_n, C_n(0), n = 1 \ldots N] \qquad (3.64)$$

where

$$r(\xi, 0) = \frac{b(\xi, 0)}{a(\xi, 0)} \qquad (3.65)$$

is the reflection coefficient, while

$$C_n(0) = \frac{b_n(0)}{a'_n(0)} \qquad (3.66)$$

is the normalization constant for the nth bound state, and

$$a'_n(0) = \left(\frac{\partial a}{\partial \zeta} \right) (\zeta_n, 0) \qquad (3.67)$$

Since there is a one-to-one correspondence between the scattering data and the potential functions, one can calculate the time evolution of Σ and then invert the transform to obtain the field $q(x, t)$. From (3.46) the time evolution of the scattering is given by

$$r(\xi, t) = r(\xi, 0)e^{-2i\xi^2 t} \tag{3.68}$$

and

$$\zeta_n(t) = \zeta_n(0)e^{-2i\zeta_n^2 t} \tag{3.69}$$

Finally, the field $q(x, t)$ is obtained. First, divide equation (3.60) by $a(\xi)(\xi - \zeta)$ and then take its FT with respect to ξ. Using the property $a(\zeta) \to 1$ as $\zeta \to \infty$ for $\Im(\zeta) \geq 0$, one obtains the following set of linear integral equations, also known as the Gelfand-Levitan-Marchenko (GLM) integral equations, for the eigenfunctions and the field at time t.

$$\psi_1(x, t, \zeta)e^{-i\zeta x} = \frac{1}{2\pi i} \int_{-\infty}^{\infty} \frac{r^*(\xi, t)\psi_2^*(x, t, \xi)}{\xi - \zeta} e^{-i\xi x} d\xi$$

$$+ \sum_{n=1}^{N} \frac{C_n^*(t)\psi_{n2}^*(x, t)}{\zeta_n^* - \zeta} e^{-i\zeta_n^* x} \tag{3.70}$$

$$\psi_2(x, t, \zeta)e^{-i\zeta x} = 1 - \frac{1}{2\pi i} \int_{-\infty}^{\infty} \frac{r^*(\xi', t)\psi_1^*(x, t, \xi)}{\xi - \zeta} e^{-i\xi x} d\xi$$

$$- \sum_{n=1}^{N} \frac{C_n^*(t)\psi_{n1}^*(x, t)}{\zeta_n^* - \zeta} e^{-i\zeta_n^* x} \tag{3.71}$$

$$q(x, t) = \frac{i}{\pi} \int_{-\infty}^{\infty} \frac{r^*(\xi, t)\psi_2^*(x, t, \xi)}{\xi - \zeta} e^{-i\xi x} d\xi$$

$$- 2 \sum_{n=1}^{N} C_n^*(t)\psi_{n2}^*(x, t)e^{-i\zeta_n^* x} \tag{3.72}$$

Here, $\psi_{nm} = \psi_m(x, t, \zeta_n)$ for $m = 1, 2$ and $n = 1, 2, \ldots N$. Equation (3.72) shows that the general solution for $q(x, t)$ is expressed as the sum of the N-soliton (second term) and the nonsoliton part that is due to the first term, which decays in time as dispersive waves. Also known as *soliton radiation*, this is commonly referred to as *ripples* in fluid dynamics.

3.3.1 1-Soliton Solution

In the particular case of an exact N-soliton solution of the Kerr law, one obtains analytic expressions of the field $q(x, t)$ from the general formula (3.72). In fact, in this case, the scattering data $r(\xi)$ vanishes, and $a(\zeta)$ in (3.61) reads

$$a(\zeta) = \prod_{n=1}^{N} \frac{\zeta - \zeta_n}{\zeta - \zeta_n^*} \tag{3.73}$$

where N represents the number of discrete eigenvalues. With $r(\xi) = 0$, the GLM equations (3.70) and (3.71) reduce to the case of integral equations with degenerate kernels and hence formulate algebraic equations for the column vectors

$$F_l = \begin{pmatrix} f_{l1} \\ f_{l2} \\ \cdots \\ f_{lN} \end{pmatrix} \tag{3.74}$$

where $f_{ln} = \sqrt{C_n}\psi_l(x;\zeta_n)$ for $l = 1, 2$,

$$(I + M^*M)F_2^* = E^* \tag{3.75}$$

and

$$F_1 = -MF_2^* \tag{3.76}$$

where I is the $N \times N$ identity matrix, and the $N \times N$ matrix M is defined as

$$M_{nm} = \frac{e_n e_m^*}{\zeta_n - \zeta_m^*} \tag{3.77}$$

with

$$e_n = \sqrt{C_n}e^{i\zeta_n x} \tag{3.78}$$

while E represents the column vector

$$E = \begin{pmatrix} e_1 \\ e_2 \\ \cdots \\ e_n \end{pmatrix} \tag{3.79}$$

From (3.72) the N-soliton solution can be written as

$$q(x, t) = -2\sum_{n=1}^{N} e_n^* f_{2n}^* \tag{3.80}$$

In the case of a 1-soliton solution, one obtains

$$\psi_{11}(x, t) = \frac{(\zeta_1^* - \zeta_1)C_1^*(t)e^{-i\zeta_1^* x}}{(\zeta_1^* - \zeta_1)^2 e^{i(\zeta_1^* - \zeta_1)x} - |C_1(t)|^2 e^{-i(\zeta_1^* - \zeta_1)x}} \tag{3.81}$$

$$\psi_{12}(x, t) = \frac{(\zeta_1^* - \zeta_1)^2 e^{-i\zeta_1^* x}}{(\zeta_1^* - \zeta_1)^2 e^{i(\zeta_1^* - \zeta_1)x} - |C_1(t)|^2 e^{-i(\zeta_1^* - \zeta_1)x}} \tag{3.82}$$

$$q(x, t) = -2C_1^*(x)\frac{(\zeta_1^* - \zeta_1)^2 e^{i(\zeta_1^* + \zeta_1)x}}{(\zeta_1^* - \zeta_1)^2 e^{-i(\zeta_1 - \zeta_1^*)x} - |C_1(t)|^2 e^{-i(\zeta_1 - \zeta_1^*)x}} \tag{3.83}$$

One can now express the scattering data as

$$\zeta_1 = \frac{1}{2}(\kappa + iA) \tag{3.84}$$

$$C_1(t) = -Ae^{\sigma_1(t) - i\theta_1(t)} \tag{3.85}$$

where $\sigma_1(t) = A\kappa t + \sigma_{10}$ and $\theta_1(t) = (A^2 - \kappa^2)t/2 + \theta_{10}$. One obtains from (3.83) the 1-soliton solution to (3.34) as given by (3.40). On comparing with the notation that was introduced in chapter 2, the function g in (2.36) for Kerr law is therefore given by

$$g[B(x - \bar{x}(t))] = \frac{1}{\cosh[B(x - \bar{x}(t))]} = \frac{1}{\cosh \tau} \tag{3.86}$$

where κ is the frequency of the soliton, ω is the wave number, and σ_0 is the center of phase of the soliton. Also, the corresponding parameter dynamics are

$$\frac{dA}{dt} = 0 \tag{3.87}$$

$$\frac{dB}{dt} = 0 \tag{3.88}$$

$$\frac{d\kappa}{dt} = 0 \tag{3.89}$$

and

$$\frac{d\bar{x}}{dt} = -\kappa \tag{3.90}$$

3.4 Integrals of Motion

An important property of the NLSE with the Kerr law of nonlinearity is that it has an infinite number of integrals of motion. This characteristic is closely related to the properties of the soliton solutions that are seen in the previous section. In fact, each conserved quantity gives a restriction, or control, on the evolution of the solitons. By Noether's theorem, each conserved law corresponds to a symmetry that leaves the NLSE invariant. Examples of these symmetries are the translational invariance and phase invariance.

Basically, in order to derive these conserved quantities, one exploits the fact that the inverse transmission coefficient $a(\zeta)$ in equation (3.73) is invariant, namely $\partial a/\partial t = 0$. As noted in the previous section, whenever $|\zeta| \gg 1$ and $\Im(\zeta) \geq 0$, the function $\ln a(\zeta)$ is analytic on the upper half plane $\Im(\zeta) \geq 0$ and $\ln a(\zeta) \to 0$ as $|\zeta| \to \infty$. Therefore, one may write the expansion [5]

$$\ln a(\zeta) = \sum_{n=1}^{\infty} \frac{C_n}{\zeta^n} \tag{3.91}$$

By using (3.73) (for $\Im(\zeta) > 0$ and $x \to \infty$, where $\psi \to 0$) and equation (3.84), one can obtain the conserved quantities of the NLSE as follows

$$E = \int_{-\infty}^{\infty} |q|^2 dx = 2A \tag{3.92}$$

$$M = \frac{i}{2} \int_{-\infty}^{\infty} (qq_x^* - q^*q_x)dx = -2\kappa A \tag{3.93}$$

$$H = \frac{1}{2} \int_{-\infty}^{\infty} (|q_x|^2 - |q|^4)dx = \frac{2}{3}A(3\kappa^2 - A^2) \tag{3.94}$$

The values of the energy, momentum, and the Hamiltonian are obtained by using the 1-soliton solution of the NLSE given in (3.40). One additional conserved quantity is

$$H_1 = \frac{1}{2} \int_{-\infty}^{\infty} \{q_{xx}q_x^* - q_x q_{xx}^* + 3|q|^2(q^*q_x - qq_x^*)\}dx = 2\kappa A(\kappa^2 - A^2) \tag{3.95}$$

which represents the next hierarchy of the NLSE.

For the perturbed NLSE with Kerr law given by

$$iq_t + \frac{1}{2}q_{xx} + |q|^2 q = i\epsilon R[q, q^*] \tag{3.96}$$

the adiabatic parameter dynamics of the solitons are governed by

$$\frac{dA}{dt} = \frac{dB}{dt} = \frac{\epsilon}{2} \int_{-\infty}^{\infty} (q^*R + qR^*)dx \tag{3.97}$$

and

$$\frac{d\kappa}{dt} = \frac{\epsilon}{2A} \left[i \int_{-\infty}^{\infty} (q_x^*R - q_x R^*)dx - \kappa \int_{-\infty}^{\infty} (q^*R + qR^*)dx \right] \tag{3.98}$$

which can be obtained from (2.51) and (2.54), respectively, since $E = 2A$ for Kerr law nonlinearity. For Kerr law, with

$$R = \delta|q|^{2m}q + \sigma q \int_{-\infty}^{x} |q|^2 ds \tag{3.99}$$

equations (2.51) and (2.52) reduce to

$$\frac{dE}{dt} = 2\epsilon\delta A^{2m+1} \frac{\Gamma(\frac{1}{2})\Gamma(m+1)}{\Gamma(m+\frac{3}{2})} + 4\epsilon\sigma A^2 \tag{3.100}$$

$$\frac{dM}{dt} = 2\epsilon\delta\kappa A^{2m+1} \frac{\Gamma(\frac{1}{2})\Gamma(m+1)}{\Gamma(m+\frac{3}{2})} + 4\epsilon\sigma\kappa A^2 \tag{3.101}$$

Note that for the case of Kerr law, $A = B$. Considering the particular perturbation terms in (3.99), equations (3.97) and (3.98) integrate, on using the

1-soliton solution given in (3.40) to

$$\frac{dA}{dt} = \frac{dB}{dt} = \epsilon\delta A^{2m+1}\frac{\Gamma'(\frac{1}{2})\Gamma(m+1)}{\Gamma(m+\frac{3}{2})} + 2\epsilon\sigma A^2 \tag{3.102}$$

$$\frac{d\kappa}{dt} = 0 \tag{3.103}$$

Finally, to obtain the velocity of the soliton for Kerr law, use equation (2.55) and the perturbation terms given by (3.99). This integrates out to

$$v = \frac{d\bar{x}}{dt} = -\kappa + \frac{\epsilon}{E}\int_{-\infty}^{\infty} x(q^*R + qR^*)dx = -\kappa + \epsilon\sigma \tag{3.104}$$

where the energy E is given by (3.92).

3.4.1 Hamiltonian Structure

It is known that equations solvable by the IST, such as the NLSE with Kerr law nonlinearity, are completely integrable Hamiltonian systems and the IST amounts to a canonical transformation from physical variables to an infinite set of action-angle variables. The phase space M_0 is an infinite dimensional real linear space with complex coordinates defined by a pair of functions $q(x), r(x)$ for which the variable x may be thought of as a coordinate label [153].

On the algebra of smooth functionals on the phase space M_0, a Poisson structure is defined by the following Poisson brackets

$$\{F, G\} = i\int_{-\infty}^{\infty}\left(\frac{\delta F}{\delta q(x)}\frac{\delta G}{\delta r(x)} - \frac{\delta F}{\delta r(x)}\frac{\delta G}{\delta q(x)}\right)dx \tag{3.105}$$

where the variational derivative is defined as

$$\delta F(q, r) = F(q + \delta q, r + \delta r) - F(q, r)$$

$$= \int_{-\infty}^{\infty}\left(\frac{\delta F}{\delta q(x)}\delta q(x) + \frac{\delta F}{\delta r(x)}\delta r(x)\right)dx \tag{3.106}$$

The bracket in (3.105) possesses the basic properties of Poisson brackets, namely

1. skew-symmetry: $\{F, G\} = -\{G, F\}$
2. linearity: $\{aF + bG, H\} = a\{F, H\} + b\{G, H\}$ for any constants a and b
3. Jacobi identity: $\{F, \{G, H\}\} + \{H, \{F, G\}\} + \{G, \{H, F\}\} = 0$

It is to be noted that the bracket in (3.105) is the infinite dimensional generalization of the usual Poisson bracket in the phase space R^{2n} with real coordinates

p_k, q_k for $1 \leq k \leq n$, namely

$$\{f, g\} = \sum_{k=1}^{n} \left(\frac{\partial f}{\partial q_k} \frac{\partial g}{\partial p_k} - \frac{\partial f}{\partial p_k} \frac{\partial g}{\partial q_k} \right) \tag{3.107}$$

The coordinates $q(x), r(x)$ themselves may be considered as functional on M_0. Their variational derivatives are the generalized functions

$$\frac{\delta q(x)}{\delta q(y)} = \delta(x - y) \tag{3.108}$$

$$\frac{\delta r(x)}{\delta q(y)} = \delta(x - y) \tag{3.109}$$

and

$$\frac{\delta q(x)}{\delta r(y)} = \frac{\delta r(x)}{\delta q(y)} = \delta(x - y) \tag{3.110}$$

where $\delta(x - y)$ is the Dirac δ-function. Substituting (3.108) and (3.109) into (3.105) gives

$$\{q(x), q(y)\} = \{r(x), r(y)\} = 0 \tag{3.111}$$

$$\{q(x), r(y)\} = i\delta(x - y) \tag{3.112}$$

These results also give

$$\frac{\delta F}{\delta q(x)} = -i\{F, r(x)\} \tag{3.113}$$

$$\frac{\delta F}{\delta r(x)} = -i\{F, q(x)\} \tag{3.114}$$

A dynamical system is said to be Hamiltonian if one can identify generalized coordinates q, momenta p, and a Hamiltonian $H(p, q, t)$ such that the equations of motion of the system can be written as

$$\frac{\partial q}{\partial t} = \{q, H\} \tag{3.115}$$

$$\frac{\partial p}{\partial t} = \{p, H\} \tag{3.116}$$

where $\{,\}$ denotes the Poisson bracket. Equations (3.115) and (3.116) are called *Hamilton's equations of motion*, and the variables (p, q) are called the *conjugates*.

3.5 Variational Principle

Linear evolution equations that are studied in mathematical physics have a variety of analytical techniques to solve them. Most of these methods—namely the Fourier series method, integral transform techniques, and Green's function techniques—make direct or indirect use of the superposition principle. However, for nonlinear evolution equations, the superposition technique does not hold and most of the analytical solution techniques become obsolete. Although numerical experiments give an understanding of the behavior of these equations, they serve as inspiration and guidance for theoretical efforts, and significant efforts were subsequently made in the analytical techniques. One such outstanding technique is the IST previously discussed. However, it should be emphasized that the number of nonlinear equations that can be solved by the IST is very limited, for example, to the NLSE with Kerr law nonlinearity, the derivative NLSE, the Korteweg–de Vries equation, the Sin-Gordon equation, and a few others. Furthermore, even in situations when the method can be applied, the explicit information that can be extracted from the solution is rather limited. This situation has prompted an effort to complement the exact analytical solution methods by approximate methods that sacrifice exactness in order to obtain explicit results and a clear physical picture of the properties of the solution. One such method that has been found very useful in many investigations in nonlinear optics is a direct variational method based on trial functions and Rayleigh-Ritz optimization. This method has several advantages for studying solitons in the nonlinear optics community. Some of these advantages are [38]

1. This method is applicable to a perturbation problem for which the unperturbed system may not be integrable. Also, this method only requires that the unperturbed system admits a well-defined solution, such as a soliton, although this method has limitations.

2. It is a universal method that is suitable for equations in any dimensions with external forces and potentials.

3. It often gives results that are quite similar to numerical simulations.

4. The method is simple and can be learned in a very short time.

5. This approach leads to a simplified, finite-dimensional dynamical system with interesting and rich properties. Consequently, it is easier to study the reduced system than the original problem.

For a finite dimensional problem of a single particle, the temporal development of its position is given by Hamilton's principle of least action, which states that the action given by the time integral of the Lagrangian is an extremum, namely

$$\delta \int_{t_1}^{t_2} L(x, \dot{x}) dt = 0 \tag{3.117}$$

where x is the position of the particle and $\dot{x} = dx/dt$. The variational problem (3.117) then leads to the familiar Euler-Lagrange (EL) equation

$$\frac{\partial L}{\partial x} - \frac{d}{dt}\left(\frac{\partial L}{\partial \dot{x}}\right) = 0 \qquad (3.118)$$

The derivation of this EL equation can be found in the familiar work by Goldstein. Here, it is assumed that the variations of x and $\dot{x}$, namely δx and $\delta \dot{x} = d(\delta x)/dt$, vanish at the boundary of the integration. To obtain the Hamiltonian formulation of (3.118), one needs to define the canonical momentum p to the position x as

$$p = \frac{\partial L}{\partial \dot{x}} \qquad (3.119)$$

The Hamiltonian H is defined by the Legendre transformation

$$H = p\dot{x} - L(x, \dot{x}) \qquad (3.120)$$

so that (3.119) and (3.120) lead to the Hamilton's equations

$$\frac{dx}{dt} = \frac{\partial H}{\partial p} \qquad (3.121)$$

and

$$\frac{dp}{dt} = -\frac{\partial H}{\partial x} \qquad (3.122)$$

For the case of an infinite dimensional problem, such as the NLSE with Kerr law of nonlinearity, Hamilton's principle is extended to the extremum of the action given by the integral of the Lagrangian, which is a real function of the fields q, q^* and their derivatives

$$\delta \int \int L(q, q^*, q_t, q_t^*, q_x, q_x^*, q_{xx}, \ldots) dx\,dt = 0 \qquad (3.123)$$

where L is given by

$$L = \frac{1}{2} \int_{-\infty}^{\infty} [i(q^*q_t - qq_t^*) + |q|^4 - |q_x|^2] dx \qquad (3.124)$$

The variation of (3.124) is then defined as

$$\delta \int \int L(q, q^*, \ldots) dx\,dt$$

$$= \lim_{\epsilon \to 0} \frac{1}{\epsilon} \int \int \{L(q + \epsilon(\delta q), q^* + \epsilon(\delta q^*), \ldots) - L(q, q^*, \ldots)\} dx\,dt \qquad (3.125)$$

where the variations δq and δq^* are assumed to vanish at the boundary of the integration. One has, as usual, the definitions $\delta q_t = \partial(\delta q)/\partial t$, $\delta q_t^* = \partial(\delta q^*)/\partial t$,

$\delta q_x = \partial(\delta q)/\partial x$, and so on. Then, after integration by parts, (3.125) reduces to

$$\int \int \left[\left\{ \sum_{n=0}^{\infty} (-1)^n \frac{\partial^n}{\partial x^n} \frac{\partial L}{\partial q_{nx}^*} - \frac{\partial}{\partial t} \frac{\partial L}{\partial q_t^*} \right\} \delta q^* + \{cc\} \right] dx dt = 0 \qquad (3.126)$$

where $q_{nx}^* = \partial^n q^*/\partial x^n$ and cc represents the complex conjugate, while it is to be noted that $(\partial L/\partial q_{nx})^* = \partial L/\partial q_{nx}^*$. Since the variations δq and δq^* are taken to be arbitrary and independent, one has

$$\sum_{n=0}^{\infty} (-1)^n \frac{\partial^n}{\partial x^n} \frac{\partial L}{\partial q_{nx}^*} - \frac{\partial}{\partial t} \frac{\partial L}{\partial q_t^*} = 0 \qquad (3.127)$$

and its complex conjugate. The function that makes the variational functional stationary is also a solution of the corresponding nonlinear evolution equation. In this process, an intelligent guess is made for the evolution of $q(x, t)$ in the sense that the form of q as a function of x is modeled in terms of certain parameter functions, l, that characterize the crucial features of the solutions, namely the amplitude, spatial width, phase variations, and others. The parameters of this trial function are allowed to be functions of time (i.e. $l = l(t)$). Inserting the trial function into the variational integral, the spatial integration can be performed and a reduced variational problem for the parameter functions $l(t)$ is obtained. The EL equation of the reduced variational problem becomes

$$\frac{\partial L}{\partial l} - \frac{d}{dt} \left(\frac{\partial L}{\partial l_t} \right) = 0 \qquad (3.128)$$

On using the ansatz for the soliton solution of the NLSE [185]

$$q(x, t) = \frac{A(t)}{\cosh [B(t)(x - \bar{x}(t))]} e^{i\{-\kappa(t)(x - \bar{x}(t)) + \delta(t)\}} \qquad (3.129)$$

the Lagrangian that is given by (3.124) reduces to

$$L = -2A \left(\kappa \frac{d\bar{x}}{dt} + \frac{d\delta}{dt} \right) + \frac{1}{3} A^3 - A\kappa^2 \qquad (3.130)$$

Substituting $A, B, \kappa, \bar{x}$, and δ for l in (3.128) yields the following set of equations

$$\frac{dA}{dt} = 0 \qquad (3.131)$$

$$\frac{dB}{dt} = 0 \qquad (3.132)$$

$$\frac{d\kappa}{dt} = 0 \qquad (3.133)$$

$$\frac{d\bar{x}}{dt} = -\kappa \qquad (3.134)$$

and

$$\frac{d\delta}{dt} = \frac{1}{2}(A^2 + \kappa^2) \tag{3.135}$$

In the presence of the perturbation terms, the EL equation for (3.128) can be extended to

$$\frac{\partial L}{\partial l} - \frac{d}{dt}\left(\frac{\partial L}{\partial l_t}\right) = i\epsilon \int_{-\infty}^{\infty} \left(R\frac{\partial q^*}{\partial l} - R^*\frac{\partial q}{\partial l}\right) dx \tag{3.136}$$

where l represents the five soliton parameters. Once again, substituting A, B, κ, $\bar{x}$, and δ for l in (3.136), the following adiabatic evolution equations are obtained

$$\frac{dA}{dt} = \frac{dB}{dt} = \epsilon \int_{-\infty}^{\infty} \Re[Re^{-i\phi}]\frac{1}{\cosh\tau}d\tau \tag{3.137}$$

$$\frac{d\kappa}{dt} = -\epsilon \int_{-\infty}^{\infty} \Im[Re^{-i\phi}]\frac{\tanh\tau}{\cosh 2\tau}d\tau \tag{3.138}$$

$$\frac{d\bar{x}}{dt} = -\kappa + \frac{\epsilon}{A^2} \int_{-\infty}^{\infty} \Re[Re^{-i\phi}]\frac{\tau}{\cosh\tau}d\tau \tag{3.139}$$

$$\frac{d\delta}{dt} = \frac{1}{2}(\kappa^2 + A^2) + \frac{\epsilon}{A} \int_{-\infty}^{\infty} \Im[Re^{-i\phi}]\frac{1}{\cosh\tau}(1 - \tanh\tau)d\tau$$

$$+ \epsilon\frac{\kappa}{A^2} \int_{-\infty}^{\infty} \Re[Re^{-i\phi}]\frac{\tau}{\cosh\tau}d\tau \tag{3.140}$$

where, once again

$$\tau = B(t)(x - \bar{x}(t))$$

while

$$\phi = -\kappa(t)(x - \bar{x}(t)) - \delta(t)$$

Equations (3.137)–(3.140) give the adiabatic dynamics of the soliton parameters in the presence of perturbation terms.

3.6 Quasi-Stationary Solution

In this section, in order to obtain the quasi-stationary (QS) solution to (3.96) for R given by (3.99), the technique that was developed in section 2.5.1 will be followed. So, using the ansatz given by (2.76), and keeping in mind that $F(s) = s$, equation (2.77) reduces to

$$-\left\{\rho_T^{(0)} + \frac{1}{2}(\rho_X^{(0)})^2\right\}\hat{q}^{(0)} + \frac{1}{2}\frac{\partial^2\hat{q}^{(0)}}{\partial\theta^2} + (\hat{q}^{(0)})^3 = 0 \tag{3.141}$$

and

$$(\rho_X^{(0)} - v^{(0)})\frac{\partial \hat{q}^{(0)}}{\partial \theta} = 0 \tag{3.142}$$

Now, (3.142) implies

$$\rho_X^{(0)} = v^{(0)} \tag{3.143}$$

Setting

$$\frac{B^2}{2} = \rho_T^{(0)} + \frac{1}{2}(\rho_X^{(0)})^2 = \rho_T^{(0)} + \frac{1}{2}(v^{(0)})^2 \tag{3.144}$$

equation (3.141) changes to

$$-\frac{B^2}{2}\hat{q}^{(0)} + \frac{1}{2}\frac{\partial^2 \hat{q}^{(0)}}{\partial \theta^2} + (\hat{q}^{(0)})^3 = 0 \tag{3.145}$$

whose solution is

$$\hat{q}^{(0)} = \frac{A}{\cosh \tau} \tag{3.146}$$

where

$$\tau = B(\theta - \bar{\theta}) \tag{3.147}$$

while

$$A = B \tag{3.148}$$

and

$$\frac{d\bar{\theta}}{dt} = v \tag{3.149}$$

At $O(\epsilon)$ level, decomposing $\hat{q}^{(1)} = \hat{\phi}^{(1)} + i\hat{\psi}^{(1)}$ yields

$$-\frac{B^2}{2}\hat{\phi}^{(1)} + \frac{1}{2}\frac{\partial^2 \hat{\phi}^{(1)}}{\partial \theta^2} + 3(\hat{q}^{(0)})^2\hat{\phi}^{(1)} = \{\rho_T^{(1)} + v^{(0)}\rho_X^{(1)}\}\hat{q}^{(0)} - \frac{\partial^2 \hat{q}^{(0)}}{\partial \theta \partial X} \tag{3.150}$$

and

$$-\frac{B^2}{2}\hat{\psi}^{(1)} + \frac{1}{2}\frac{\partial^2 \hat{\psi}^{(1)}}{\partial \theta^2} + (\hat{q}^{(0)})^2\hat{\psi}^{(1)} = -\frac{\partial \hat{q}^{(0)}}{\partial T} - v^{(0)}\frac{\partial \hat{q}^{(0)}}{\partial X}$$

$$-\{\rho_X^{(1)} - v^{(1)} + \sigma\}\frac{\partial \hat{q}^{(0)}}{\partial \theta}$$

$$- \rho_{XX}^{(0)}\hat{q}^{(0)} - \delta(\hat{q}^{(0)})^{2m+1} + \sigma\hat{q}^{(0)}\int_{-\infty}^{x}(\hat{q}^{(0)})^2 ds \tag{3.151}$$

Here, as discussed in the previous chapter, setting $\rho_{XX}^{(0)} = 0$ in (3.151) to eliminate frequency chirp gives

$$-\frac{B^2}{2}\hat{\psi}^{(1)} + \frac{1}{2}\frac{\partial^2 \hat{\psi}^{(1)}}{\partial \theta^2} + (\hat{q}^{(0)})^2 \hat{\psi}^{(1)} = -\frac{\partial \hat{q}^{(0)}}{\partial T} - v^{(0)}\frac{\partial \hat{q}^{(0)}}{\partial X}$$

$$-\{\rho_X^{(1)} - v^{(1)} + \sigma\}\frac{\partial \hat{q}^{(0)}}{\partial \theta}$$

$$- \delta(\hat{q}^{(0)})^{2m+1} + \sigma\hat{q}^{(0)}\int_{-\infty}^{x}(\hat{q}^{(0)})^2 ds \qquad (3.152)$$

Fredholm's alternative, when applied to (3.151), yields

$$\frac{\partial B}{\partial X} = 0 \qquad (3.153)$$

and

$$\rho_T^{(1)} + v^{(0)}\rho_X^{(1)} = 0 \qquad (3.154)$$

whereas applied to (3.152) yields

$$\frac{dB}{dT} = \delta B^{2m+1}\frac{\Gamma\left(\frac{1}{2}\right)\Gamma(m+1)}{\Gamma\left(m+\frac{3}{2}\right)} + 2\sigma B^2 \qquad (3.155)$$

and

$$\rho_X^{(1)} = v^{(1)} - \sigma \qquad (3.156)$$

Equation (3.153) shows that B is a function of T only and so is A, since $A = B$ for Kerr law. Thus, these $O(\epsilon)$ equations reduce to

$$-\frac{B^2}{2}\hat{\phi}^{(1)} + \frac{1}{2}\frac{\partial^2 \hat{\phi}^{(1)}}{\partial \theta^2} + 3(\hat{q}^{(0)})^2\hat{\phi}^{(1)} = 0 \qquad (3.157)$$

and

$$-\frac{B^2}{2}\hat{\psi}^{(1)} + \frac{1}{2}\frac{\partial^2 \hat{\psi}^{(1)}}{\partial \theta^2} + (\hat{q}^{(0)})^2\hat{\psi}^{(1)} = -\frac{\partial \hat{q}^{(0)}}{\partial T} - \delta(\hat{q}^{(0)})^{2m+1}$$

$$+ \sigma\hat{q}^{(0)}\int_{-\infty}^{x}(\hat{q}^{(0)})^2 ds \qquad (3.158)$$

whose solutions are respectively

$$\hat{\phi}^{(1)} = 0 \qquad (3.159)$$

and

$$\hat{\psi}^{(1)} = \frac{\partial\bar{\theta}}{\partial T}\frac{\tau}{\cosh \tau} - \frac{1}{2B^2}\frac{dB}{dT}\left(\frac{\tau^2}{\cosh \tau} + \cosh \tau\right)$$

$$+ \sigma \cosh \tau + 2\delta\frac{B^{2m-1}}{\cosh \tau}\int^{\tau}\cosh^2 s_2 \left(\int^{s_2}\frac{1}{\cosh^{2m+2} s_1}ds_1\right)ds_2 \qquad (3.160)$$

which leads to the QS solution (2.98) for the Kerr law of nonlinearity.

Now, the dynamical system in (3.102) and (3.103) has a stable, fixed point, called a *sink*, that is given by $(A, \kappa) = (\bar{A}, 0)$ or $(B, \kappa) = (\bar{B}, 0)$, where

$$\bar{A} = \bar{B} = \left[\frac{2\sigma}{\delta} \frac{\Gamma(m + \frac{3}{2})}{\Gamma(\frac{1}{2})\Gamma(m+1)} \right]^{\frac{1}{2m-1}} \qquad (3.161)$$

Thus, there exists a steady-state soliton whose amplitude and frequency (velocity) get locked to this fixed value in a medium where the energy growth rate and energy losses fully compensate for each other. The QS soliton given by (2.98) for Kerr law travels with the velocity given by (3.104) through an optical fiber with a fixed amplitude and width that are given by (3.161).

3.7 Lie Transform

In this section, the lie transform (LT) technique will be used to analyze the perturbed NLSE that contains only Hamiltonian-type perturbations. This is because the LT technique is restricted to dispersive-type perturbations. So, equation (3.96) will be studied with R given by [83, 185, 241]

$$R = \lambda(|q|^2 q)_x + \mu(|q|^2)_x q - \gamma q_{xxx} \qquad (3.162)$$

Thus, the perturbed NLSE for Kerr law nonlinearity that will be studied using the LT is

$$iq_t + \frac{1}{2}q_{xx} + |q|^2 q = i\epsilon\{\lambda(|q|^2 q)_x + \mu(|q|^2)_x q - \gamma q_{xxx}\} \qquad (3.163)$$

In (3.163), λ is the self-steepening coefficient for short pulses (typically ≤ 100 femto seconds), μ is the nonlinear dispersion coefficient, and γ is the coefficient of the third order dispersion [83, 185]. The self-steepening term creates an optical shock on the trailing edge of the pulse in the absence of group velocity dispersion effects. This phenomenon is due to the intensity dependence of the group velocity that makes the peak of the pulse move slower than the wings. The group velocity dissipates the shock and smooths the trailing edge considerably. However, self-steepening still manifests through a shift of the pulse center.

The NLSE, as given by (3.34), does not give correct prediction for pulse widths smaller than 1 picosecond. For example, in solid-state solitary lasers, where pulses as short as 10 femtoseconds are generated, the approximation breaks down. Thus, quasi-monochromaticity is no longer valid and, consequently, higher order dispersion terms creep in. If group velocity dispersion is close to zero, one needs to consider the third-order dispersion for performance enhancement along transoceanic distances. Also, for short pulse widths where the group velocity dispersion changes within the spectral bandwidth of the signal, one needs to take into account the presence of third-order dispersion.

The purpose of the LT is to transform the perturbed NLSE given by (3.163) to a simple one, commonly called the normal form of the original equation, that is integrable by the IST method. In the following subsection, the normal form theory will be introduced and, finally, the LT technique will be utilized to integrate it.

3.7.1   Introduction

In this subsection, first, the definition of the infinite-dimensional vector spaces of differential polynomials of $(q, q^*, q_x, q_x^*, q_{xx}, \ldots)$ will be given, and then several linear operators in these spaces will be given. The LT and the Hamiltonian formalism will then be defined. The degree, Deg, of the differential monomial, say $X[q, q^*]$, of $(q, q^*, q_x, q_x^*, \ldots)$ is as follows

$$\text{Deg}(X) = (\# \text{ of } q's \text{ in } X) + (\# \text{ of } q^*s \text{ in } X)$$

$$+ (\# \text{ of derivatives } \partial/\partial x \text{ in } X) \tag{3.164}$$

For example,

$$Deg(|q|^2 q_x) = Deg(q_{xxx}) = 4$$

Now, let $\chi_N[[q, q^*]]$ be the set of all differential polynomials of degree N that satisfy the symmetry property $X[e^{i\theta}q, e^{-i\theta}q^*] = e^{i\theta} X[q, q^*]$ for any real number θ. In addition, let $\chi_N^*[[q, q^*]]$ be the complex conjugate of $\chi_N[[q, q^*]]$. For example, $\chi_3[[q, q^*]] = q_{xx}, |q|^2 q$. These spaces form a finite dimensional vector space over the field of complex numbers. Define the infinite dimensional space of differential polynomials $\chi[[q, q^*]]$ as the following direct sum of the vector spaces χ_N

$$\chi[[q, q^*]] = \bigoplus_{N=1}^{\infty} \chi_N[[q, q^*]] = \{q, q_x, q_{xx}, |q|^2 q, \ldots\} \tag{3.165}$$

The space $\chi^*[[q, q^*]]$ denotes the complex conjugate of $\chi[[q, q^*]]$. In this space, define the derivative as

$$\frac{d}{dx} = \sum_{n=0}^{\infty} \left\{ q_{(n+1)x} \frac{\partial}{\partial q_{nx}} + q_{(n+1)x}^* \frac{\partial}{\partial q_{nx}^*} \right\} \tag{3.166}$$

where $q_{nx} = \partial^n q/\partial x^n$. The directional derivative, or the Lie derivative, which is denoted by $\psi \cdot \nabla$ with respect to $\psi = \psi[q, q^*] \in (\chi \times \chi^*)[[q, q^*]]$ is given by

$$\psi \cdot \nabla = \sum_{n=0}^{\infty} \left\{ \psi_{nx} \frac{\partial}{\partial q_{nx}} + \psi_{nx}^* \frac{\partial}{\partial q_{nx}^*} \right\} \tag{3.167}$$

where $\psi_{nx} = \partial^n \psi / \partial x^n$. Note that $d/dx = q_x \cdot \nabla$ so that, in view of (3.167), one can define the derivative d/dt as

$$\frac{\tilde{d}}{dt} = \left(\frac{dq}{dt}\right) \cdot \nabla \tag{3.168}$$

where t is a variable in the function $q(x, t)$ while

$$\frac{dq}{dt} = X[q, q^*] \in \chi[[q, q^*]] \tag{3.169}$$

for example, $X[q, q^*] = (i/2)q_{xx} + i|q|^2 q$. The Lie bracket $[X, Y]$ for X and $Y \in \chi[[q, q^*]]$ is defined as

$$[X, Y] = X \cdot \nabla Y - Y \cdot \nabla X \tag{3.170}$$

With (3.170), the space $\chi[[q, q^*]]$ forms an infinite dimensional Lie algebra, namely the space χ is closed under the product of (3.170).

Also, define the space $\chi^{(0)}[[q, q^*]]$ as the set of polynomials that satisfy the relation $X[e^{i\theta}q, e^{-i\theta}q^*] = e^{i\theta} X[q, q^*]$. The conserved quantities of the NLSE that are given by (3.92)–(3.94) belong to $\chi^{(0)}[[q, q^*]]$. Now, define the LT of the equation

$$\frac{dq}{dt} = X_\epsilon[q, q^*] = \sum_{n=0}^{\infty} \epsilon^n X_n[q, q^*] \tag{3.171}$$

where X_n is a homogenous polynomial of $(q, q^*, q_x, q_x^* \ldots)$ and $\mathrm{Deg}(X_{n+1}) = \mathrm{Deg}(X_n) + 1$. The LT or the Lie exponential transform $q = q_\epsilon[q, q^*] \in \chi[q, q^*]$ is defined as

$$q = \exp(\phi_\epsilon \cdot \nabla)Q \tag{3.172}$$

where the exponential of the derivative reads

$$\exp(\phi_\epsilon \cdot \nabla) = 1 + \phi_\epsilon \cdot \nabla + \frac{1}{2!}\phi_\epsilon \cdot \nabla(\phi_\epsilon \cdot \nabla) + \cdots \tag{3.173}$$

with $\phi_\epsilon \cdot \nabla$ is the directional derivative as defined in (3.167), 1 is the identity operator, and the function $\phi_\epsilon[Q, Q^*]$ is called the Lie-generating function. Here, the transformed equation for Q is given by

$$\frac{dQ}{dt} = Y_\epsilon[Q, Q^*] = \sum_{n=0}^{\infty} \epsilon^n Y_n[Q, Q^*] \tag{3.174}$$

where $Y_0 = X_0$ and $Y_n[Q, Q^*] \in \chi[[Q, Q^*]]$ with $\mathrm{Deg}(Y_n) = \mathrm{Deg}(X_n)$, being defined from the solvability condition for the equation of ϕ_ϵ, as will be shown below. From (3.171), (3.172), and (3.174), the following relation between X_ϵ, Y_ϵ, and ϕ_ϵ can be established

$$(e^{\phi_\epsilon \cdot \nabla} X_\epsilon[Q, Q^*]) \cdot \nabla = Y_\epsilon[Q, Q^*] \cdot \nabla e^{\phi_\epsilon \cdot \nabla} \tag{3.175}$$

This is the transformation rule for the vector fields (3.171) and (3.174), with the change of coordinates given by (3.172). With the definition of the Lie bracket given by (3.170), the relation given by (3.175) can be written as the Campbell-Baker-Hausdroff (CBH) formula

$$Y_\epsilon = \sum_{n=0}^{\infty} \frac{1}{n!} [\phi_\epsilon, [\phi_\epsilon, \dots [\phi_\epsilon, X_\epsilon] \dots]]$$

$$= X_\epsilon + [\phi_\epsilon, X_\epsilon] + \frac{1}{2!} [\phi_\epsilon, [\phi_\epsilon, X_\epsilon]] + \cdots \tag{3.176}$$

Assuming ϕ_ϵ in (3.172) has a power series in ϵ, namely

$$\phi_\epsilon [Q, Q^*] = \sum_{n=1}^{\infty} \epsilon^n \phi_n [Q, Q^*] \tag{3.177}$$

the CBH formula given by (3.176) gives the following equations for ϕ_n at each order

$$[X_0, \phi_1] = X_1 - Y_1 \tag{3.178}$$

$$[X_0, \phi_2] = X_2 + [\phi_1, X_1] + \frac{1}{2} [\phi_1, [\phi_1, X_0]] - Y_2$$

$$= X_2 - Y_2 + \frac{1}{2} [\phi_1, X_1 + Y_1] \tag{3.179}$$

$$[X_0, \phi_3] = X_3 - Y_3 + \frac{1}{3} [\phi_1, 2X_2 + Y_2] + \frac{1}{3} [\phi_2, X_1 + 2Y_1] + \frac{1}{6} [\phi_1, [\phi_1, X_1]] \tag{3.180}$$

and so on. In general, the equation for ϕ_n is of the form

$$[X_0, \phi_n] \equiv ad_{X_0} \phi_n = F_n - Y_n \tag{3.181}$$

where the linear operator $ad_{X_0} = [X_0, \cdot]$ is called the adjoint representation of X_0, while F_n is determined from $(X_1, \dots, X_n, Y_1, \dots Y_{n-1}, \phi_1, \dots, \phi_{n-1})$. For the NLSE $X_0 = 1/2 Q_{xx} + i|Q|^2 Q$

$$ad_{X_0} : \chi[[Q, Q^*]] \rightarrow \chi_{N+2}[[Q, Q^*]] \tag{3.182}$$

and the polynomial parts of the kernel of the operator ad_{X_0}, denoted by $ker(ad_{X_0})$, are

$$ker(ad_{X_0}) \cap \chi_1 = \{i Q\} \tag{3.183}$$

$$ker(ad_{X_0}) \cap \chi_2 = \{Q_x\} \tag{3.184}$$

$$ker(ad_{X_0}) \cap \chi_3 = \left\{ \frac{i}{2} Q_{xx} + i|Q|^2 Q = X_0 \right\} \tag{3.185}$$

The solution of (3.181) is then obtained by decomposing $F_n \in \chi_{n+3}[[Q, Q^*]]$ into the following direct sum

$$\{im(ad_{X_0}) \bigoplus (im(ad_{X_0}))^c\} \cap F_n \tag{3.186}$$

where $(im(ad_{X_0}))^c$ is the complement of the space $im(ad_{X_0})$. Then Y_n is chosen to be

$$Y_n = (im(ad_{X_0}))^c\} \cap F_n \tag{3.187}$$

which gives $(ad_{X_0}\phi_n) = im(ad_{X_0}) \cap F_n$ and thus ϕ_n can be obtained. In particular, in the choice of Y_n in (3.187), note that the linear part of Y_n, $\partial^{n+2}Q/\partial x^{n+2}$ is equal to that of F_n, since $im(ad_{X_0})$ contains only nonlinear terms. The linear part appearing in Y_n is a consequence of the linear resonances caused by the higher order corrections in the linear dispersion relation.

The solution space ϕ_n in (3.181) is denoted by $\xi_n[[Q, Q^*]]$, namely

$$\xi_n[[Q, Q^*]] = \{S[[Q, Q^*]] \mid ad_{X_0}S \in \chi_{n+3}\} \tag{3.188}$$

From (3.188)

$$\chi_{n+1}[[Q, Q^*]] \subset \xi_n[[Q, Q^*]] \tag{3.189}$$

In general, the nonpolynomial part of ψ_n, denoted by $\xi_n \backslash \chi_{n+1}$, cannot be completely determined because of the infinite dimensional nature of the operator ad_{X_0}. However, since the dimension of χ_{n+3} is finite, only a finite number of independent solutions of $ad_{X_0}\phi_n \in \chi_{n+3}$ is necessary. In a direct calculation for $n = 1$, one finds the nonpolynomial part of ξ_1 as

$$\xi_1 \backslash \chi_2[[Q, Q^*]] = \left\{ iQ \int_{-\infty}^{x} |Q(x)|^2 ds \right\} \tag{3.190}$$

while for $n = 2$, one obtains

$$\xi_2 \backslash \chi_3[[Q, Q^*]] = \left\{ Q \int_{-\infty}^{x} (Q_x Q^* - QQ_x^*)ds, \quad Q_x \int_{-\infty}^{x} |Q(x)|^2 ds \right\} \tag{3.191}$$

Similar results occur for higher degrees. However, in higher order problems for (3.181), F_n, depending on the previous solution ϕ_{n-1}, may not be a polynomial, and the algebra of $\xi = \oplus_{n=1}\xi_n$ becomes complicated. In many practical problems, however, an asymptotic result, including only the first few orders, might be necessary. Thus, one can assume the spaces of the Lie-generating functions to be

$$\phi_1 \in \xi_1 = \left\{ iQ_x, \quad iQ \int_{-\infty}^{x} |Q(x)|^2 ds \right\} \tag{3.192}$$

$$\phi_2 \in \xi_2 = \left\{ Q_{xx}, \quad |Q|^2 Q, \quad Q \int_{-\infty}^{x} (Q_x Q^* - QQ_x^*)ds, \quad iQ \int_{-\infty}^{x} |Q(x)|^2 ds \right\} \tag{3.193}$$

Note from (3.191) that $[\xi_1, \chi_4] \subset \chi_5$, which leads to $F_2 \in \chi_5$. Thus by virtue of (3.172), the original equation (3.171) gets Lie transformed into (3.174), which is characterized only by the operator ad_{X_0}. Equation (3.174) is called the normal form of (3.171).

3.7.2 Application

In this subsection, the theory of LT that was developed in the last section will be utilized to study (3.163). Note that one can rewrite equation (3.163) in the following form

$$q_t = X_0[q, q^*] + \epsilon X_1[q, q^*] + O(\epsilon^2) \tag{3.194}$$

where X_0 is the NLSE part and X_1 represents the Hamiltonian perturbation terms, namely

$$X_0[q, q^*] = \frac{i}{2} q_{xx} + i|q|^2 q \tag{3.195}$$

and

$$X_1[q, q^*] = \epsilon\{-\gamma q_{xxx} + \lambda(|q|^2 q)_x + \mu(|q|^2)_x q\} \tag{3.196}$$

Now look for the LT for the perturbed (3.163) to a simple system that is known as the normal form. By virtue of the previous subsection, the transformation is defined by

$$q = e^{\Phi \cdot \nabla} = Q + \epsilon\phi_1[Q, Q^*] + O(\epsilon^2) \tag{3.197}$$

where $\phi = \epsilon\phi_1 + O(\epsilon^2)$ satisfies (3.178), namely

$$[X_0, \phi_1] = X_1 - Y_1 \tag{3.198}$$

The function $Y_1 = Y_1[Q, Q^*]$ is the first-order term in the normal form

$$Q_t = X_0[Q, Q^*] + \epsilon Y_1[Q, Q^*] + O(\epsilon^2) \tag{3.199}$$

Following the procedure that is given in the previous subsection, the solution to (3.198) is

$$\phi_1[Q, Q^*] = -\frac{i\epsilon}{2}(\lambda + 6\gamma) - i\epsilon(6\gamma + 2\lambda + \mu)Q \int_{-\infty}^{x} |Q(s)|^2 ds \tag{3.200}$$

with

$$Y_1[Q, Q^*] = -\epsilon\gamma(Q_{xxx} + 6|Q|^2 Q_x) \tag{3.201}$$

The normal form given by (3.199) with (3.195) and (3.201) is known as the Schrödinger-Hirota equation and is integrable by the IST. The 1-soliton solution is given by

$$Q(x, t) = \frac{A}{\cosh [B(x - vt - x_0)]} e^{i(-\kappa x + \omega t + \sigma_0)} \tag{3.202}$$

where the velocity and the wave number of the soliton are respectively

$$v = -\kappa + \epsilon \gamma (B^2 - 3\kappa^2) \tag{3.203}$$

and

$$\omega = \frac{B^2 - \kappa^2}{2} + \epsilon \gamma \kappa (3B^2 - \kappa^2) \tag{3.204}$$

This is, thus, an alternative method of solving the perturbed NLSE with Kerr law nonlinearity up to $O(\epsilon)$, with Hamiltonian-type perturbations only. These solutions can also be recovered by the QS method.

Based on the discussions in this section, one can conclude that, in the regime of non-Hamiltonian-type perturbation, one is compelled to use the QS method since the LT fails in presence of non-Hamiltonian-type perturbation. Also, note that, in both methods, one can obtain a closed form of the perturbed soliton at the $O(\epsilon)$ level; however, in the QS method, one can also obtain the adiabatic parameter dynamics of optical solitons that are also recoverable using the soliton perturbation theory. Once again, the LT fails to do that. So the question arises, why would anyone want to use the LT? The QS method is widely used in various other types of perturbations, including the nonlocal type, while the LT has been very sparingly used since it came into existence in 1994 [241].

In this context, it needs to be pointed out that the QS method has already been extended beyond Kerr law nonlinearity, namely to power law, parabolic law, and dual-power law, as will be seen in subsequent chapters. However, the question of extending the method of LT beyond Kerr law nonlinearity still remains open at this stage.

Exercises

1. In the perturbed NLSE with Raman scattering and a nonlocal perturbation term given by

$$iq_t + \frac{1}{2}q_{xx} + |q|^2 q = \epsilon[iv(|q|^2)_x q + \chi q_t \int_{-\infty}^{x} |q|^2 d\tau]$$

prove that the adiabatic parameter dynamics of the soliton amplitude, width, and frequency are respectively given by

$$\frac{dA}{dt} = \frac{dB}{dt} = -\frac{2}{3}\epsilon \chi A^3$$

and

$$\frac{d\kappa}{dt} = \frac{2}{15}\epsilon A^2(5\kappa\chi - 4\nu A^2)$$

while the velocity is given by

$$v = -\kappa - \epsilon\chi A$$

2. Consider the NLSE with the perturbations that are given by

$$iq_t + \frac{1}{2}q_{xx} + |q|^2 q$$
$$= i\epsilon\left[\lambda(|q|^2 q)_x + \nu(|q|^2)_x q - \gamma q_{xxx} - i\beta_1(q^2 q_t^*)_x\right.$$
$$\left. -i\beta_2 q_x^2 q^* - i\beta_3 q^*(q^2)_{xx}\right]$$

Prove that for these perturbations the adiabatic parameter dynamics for the solitons are given by

$$\frac{dA}{dt} = \frac{dB}{dt} = \frac{d\kappa}{dt} = 0$$

Also prove that, in this case, the velocity of the soliton is given by

$$v = -\kappa - 3\epsilon\gamma\kappa^2 + \frac{\epsilon A^2}{3}\{2\kappa(4\beta_3 + \beta_2 - \beta_1) - (3\lambda + 2\nu + 3\gamma)\}$$

Such types of perturbations where there is no adiabatic deformation of soliton amplitude and frequency are known as *Hamiltonian* perturbations.

3. For the stable fixed point derived for quasi-stationary solitons, establish the fixed value of the amplitude given by (3.161)

4. Prove the Jacobi identity for Poisson bracket that is given after equation (3.106).

5. Obtain the traveling wave solution of the Schrodinger-Hirota equation

$$iq_t + \frac{1}{2}q_{xx} + |q|^2 q = i\epsilon\gamma(q_{xxx} + 6|q|^2 q_x)$$

Your wave number should match with the one that is given by (3.204).

4

Power Law Nonlinearity

In this chapter, the dynamics of optical solitons with power law nonlinearity, a generalization of Kerr law nonlinearity, are studied. Section 4.1 briefly discusses the physics of power law nonlinearity and the mathematical issues of the nonlinear Schrödinger's equation (NLSE) with power law nonlinearity. Section 4.2 reviews the three conserved quantities of the NLSE with power law nonlinearity. It also discusses the adiabatic parameter dynamics of the solitons with power law due to the presence of perturbation terms. Finally, Section 4.3 discusses the quasi-stationarity (QS) aspect of the perturbed NLSE with power law nonlinearity.

4.1 Introduction

Power law nonlinearity is exhibited in various materials, including semiconductors. This law also occurs in media for which higher order photon processes dominate at different intensities. Moreover, in nonlinear plasmas, the power law solves the problem of small-K condensation in weak turbulence theory. This law is also treated as a generalization to Kerr law nonlinearity. For power law, the refractive index is given by [74, 81, 86]

$$n = n_0 + n_2 |E|^{2p} \qquad (4.1)$$

where n_0 is the linear refractive index of the medium, n_2 is the higher order nonlinear coefficient, and E is the electric field of the light wave. The dimensionless form of the NLSE with power law is

$$iq_t + \frac{1}{2}q_{xx} + |q|^{2p}q = 0 \qquad (4.2)$$

Here, one needs to have $0 < p < 2$ to prevent wave collapse. In particular, it is necessary to have the restriction $p \neq 2$ to avoid the self-focusing singularity issue. This aspect of self-focusing singularity is discussed in detail in the book by Ablowitz and Segur [5]. Also, for $p = 1$ in (4.2) one recovers the NLSE with the Kerr law of nonlinearity. It is to be noted that (4.2) is not integrable by the inverse scattering transform (IST) unless $p = 1$, which was already discussed in the previous chapter. For the case where $p \neq 1$, the 1-soliton solution can be derived by the traveling wave ansatz that was introduced in chapter 2.

4.2   Traveling Wave Solution

For the case of power law nonlinearity, (2.1) reduces to

$$iq_t + \frac{1}{2}q_{xx} + |q|^{2p}q = 0 \tag{4.3}$$

so that (2.10) simplifies to

$$B^2 g'' - (\kappa^2 - 2\omega)g + 2A^{2p}g^{2p+1} = 0 \tag{4.4}$$

Multiplying (4.4) by g' and integrating yields

$$(g')^2 = \frac{2g^2}{B^2(p+1)}[(p+1)(\kappa^2 - \omega) - A^{2p}g^{2p}] \tag{4.5}$$

Separating variables and integrating yields

$$x - vt = \frac{(p+1)^{\frac{1}{2}}}{\sqrt{2}} \int \frac{dg}{g[(p+1)(\kappa^2 - \omega) - A^{2p}g^{2p}]^{\frac{1}{2}}} \tag{4.6}$$

Substituting

$$g^{2p} = \frac{(p+1)(\kappa^2 - \omega)}{A^{2p}\cosh^2\theta} \tag{4.7}$$

leads to the 1-soliton solution

$$q(x, t) = \frac{A}{\cosh^{\frac{1}{p}}[B(x - vt - \bar{x})]}e^{(-i\kappa x + i\omega t + i\sigma_0)} \tag{4.8}$$

where

$$\kappa = -v \tag{4.9}$$

and

$$\omega = \frac{B^2 - p^2\kappa^2}{2p^2} \tag{4.10}$$

with

$$B = A^p \left(\frac{2p^2}{1+p}\right)^{\frac{1}{2}} \tag{4.11}$$

Here, A is the amplitude of the soliton, B is its width, v is its velocity, κ is the soliton frequency, ω is the wave number, and $\bar{x}$ and σ_0 are the center of the soliton and the center of the soliton phase, respectively. Note that in this case the width of the soliton given by B is related to the amplitude A as seen in (4.11), unlike in the Kerr law case where $A = B$, although the relation (4.11) reduces to $A = B$ on setting $p = 1$.

The corresponding parameter dynamics for the solitons are given by

$$\frac{dA}{dt} = 0 \tag{4.12}$$

$$\frac{dB}{dt} = 0 \tag{4.13}$$

$$\frac{d\kappa}{dt} = 0 \tag{4.14}$$

and

$$\frac{d\bar{x}}{dt} = -\kappa \tag{4.15}$$

4.3 Integrals of Motion

The NLSE for power law nonlinearity has three conserved quantities, which can be obtained from (2.41)–(2.43) using the fact that $F(s) = s^p$ so that $f(s) = s^{p+1}/p + 1$. The three integrals of motion for the power law, respectively, are [81, 86]

$$E = \int_{-\infty}^{\infty} |q|^2 dx = A^{2-p} \left(\frac{1+p}{2p^2}\right)^{\frac{1}{2}} \frac{\Gamma\left(\frac{1}{2}\right)\Gamma\left(\frac{1}{p}\right)}{\Gamma\left(\frac{1}{p} + \frac{1}{2}\right)}$$

$$= B^{\frac{2-p}{p}} \left(\frac{1+p}{2p^2}\right)^{\frac{1}{p}} \frac{\Gamma\left(\frac{1}{2}\right)\Gamma\left(\frac{1}{p}\right)}{\Gamma\left(\frac{1}{p} + \frac{1}{2}\right)} \tag{4.16}$$

$$M = \frac{i}{2} \int_{-\infty}^{\infty} (q^* q_x - q q_x^*) dx$$

$$= 2\kappa A^{2-p} \left(\frac{1+p}{2p^2}\right)^{\frac{1}{p}} \frac{\Gamma\left(\frac{1}{2}\right)\Gamma\left(\frac{1}{p}\right)}{\Gamma\left(\frac{1}{p} + \frac{1}{2}\right)}$$

$$= 2\kappa B^{\frac{2-p}{p}} \left(\frac{1+p}{2p^2}\right)^{\frac{1}{p}} \frac{\Gamma\left(\frac{1}{2}\right)\Gamma\left(\frac{1}{p}\right)}{\Gamma\left(\frac{1}{p} + \frac{1}{2}\right)} \tag{4.17}$$

and

$$
\begin{aligned}
H &= \int_{-\infty}^{\infty} \left[\frac{1}{2}|q_x|^2 - \frac{1}{p+1}|q|^{2p+2} \right] dx \\
&= \frac{B^{\frac{2}{p}}}{2p^2} \left(\frac{1+p}{2p^2} \right)^{\frac{1}{p}} \left[\frac{(B^2 + \kappa^2 p^2)}{B} \frac{\Gamma\left(\frac{1}{2}\right)\Gamma\left(\frac{1}{p}\right)}{\Gamma\left(\frac{1}{p}+\frac{1}{2}\right)} - 2B \frac{\Gamma\left(\frac{1}{2}\right)\Gamma\left(\frac{p+1}{p}\right)}{\Gamma\left(\frac{p+1}{p}+\frac{1}{2}\right)} \right] \\
&= \frac{A^2}{2p^2} \left[\left\{ A^p \left(\frac{2p^2}{1+p} \right)^{\frac{1}{p}} + \frac{\kappa^2 p^2}{A^p} \left(\frac{1+p}{2p^2} \right)^{\frac{1}{2}} \right\} \frac{\Gamma\left(\frac{1}{2}\right)\Gamma\left(\frac{1}{p}\right)}{\Gamma\left(\frac{1}{p}+\frac{1}{2}\right)} \right. \\
&\quad \left. - 2A^p \left(\frac{2p^2}{1+p} \right)^{\frac{1}{2}} \frac{\Gamma\left(\frac{1}{2}\right)\Gamma\left(\frac{p+1}{p}\right)}{\Gamma\left(\frac{p+1}{p}+\frac{1}{2}\right)} \right]
\end{aligned}
\tag{4.18}
$$

In this chapter, the perturbed NLSE with power law nonlinearity that is going to be studied is

$$
i q_t + \frac{1}{2} q_{xx} + |q|^{2p} q = i\epsilon R[q, q^*]
\tag{4.19}
$$

where

$$
R = \delta |q|^{2m} q + \sigma q \int_{-\infty}^{x} |q|^2 ds
\tag{4.20}
$$

In the presence of the perturbation term in (4.20), one recovers the *modified integrals of motion*. Using the first two conserved quantities, (4.16) and (4.17), the adiabatic variation of the soliton parameters are

$$
\frac{dA}{dt} = \frac{\epsilon}{2-p} A^{p-1} \left(\frac{2p^2}{1+p} \right)^{\frac{p-1}{2p}} \frac{\Gamma\left(\frac{1}{p}+\frac{1}{2}\right)}{\Gamma\left(\frac{1}{2}\right)\Gamma\left(\frac{1}{p}\right)} \int_{-\infty}^{\infty} (q^* R + q R^*) dx
\tag{4.21}
$$

$$
\frac{dB}{dt} = \epsilon \frac{p}{2-p} B^{\frac{2p-2}{p}} \left(\frac{2p^2}{1+p} \right)^{\frac{1}{p}} \frac{\Gamma\left(\frac{1}{p}+\frac{1}{2}\right)}{\Gamma\left(\frac{1}{2}\right)\Gamma\left(\frac{1}{p}\right)} \int_{-\infty}^{\infty} (q^* R + q R^*) dx
\tag{4.22}
$$

and

$$
\begin{aligned}
\frac{d\kappa}{dt} &= \epsilon B^{\frac{p-2}{p}} \left(\frac{2p^2}{1+p} \right)^{\frac{1}{p}} \frac{\Gamma\left(\frac{1}{p}+\frac{1}{2}\right)}{\Gamma\left(\frac{1}{2}\right)\Gamma\left(\frac{1}{p}\right)} \\
&\quad \cdot \left[i \int_{-\infty}^{\infty} (q_x^* R - q_x R^*) dx - \kappa \int_{-\infty}^{\infty} (q^* R + q R^*) dx \right]
\end{aligned}
\tag{4.23}
$$

The adiabatic variations of the energy and momentum of the soliton can be obtained from equations (2.51) and (2.52), respectively, for power law nonlinearity. They are

$$\frac{dE}{dt} = 2\epsilon \left[\frac{\delta A^{2m+2}}{B} \frac{\Gamma(\frac{1}{2})\Gamma(\frac{m+1}{p})}{\Gamma(\frac{m+1}{p} + \frac{1}{2})} \right.$$

$$\left. + \frac{\sigma A^4}{B^2} \int_{-\infty}^{\infty} \frac{1}{\cosh^{\frac{2}{p}} \tau} \left(\int_{-\infty}^{\tau} \frac{1}{\cosh^{\frac{2}{p}} s} ds \right) d\tau \right] \tag{4.24}$$

and

$$\frac{dM}{dt} = 2\epsilon\kappa \left[\frac{\delta A^{2m+2}}{B} \frac{\Gamma(\frac{1}{2})\Gamma(\frac{m+1}{p})}{\Gamma(\frac{m+1}{p} + \frac{1}{2})} \right.$$

$$\left. + \frac{\sigma A^4}{B^2} \int_{-\infty}^{\infty} \frac{1}{\cosh^{\frac{2}{p}} \tau} \left(\int_{-\infty}^{\tau} \frac{1}{\cosh^{\frac{2}{p}} s} ds \right) d\tau \right] \tag{4.25}$$

Substituting the perturbation term R from (4.20) and carrying out the integrations in (4.21), (4.22), and (4.23) for the soliton given by (4.8), one obtains

$$\frac{dA}{dt} = \frac{2\epsilon\delta}{2-p} A^{2m+1} \left(\frac{1+p}{2p^2} \right)^{\frac{1}{2p}} \frac{\Gamma(\frac{1}{p} + \frac{1}{2})}{\Gamma(\frac{1}{p})} \frac{\Gamma(\frac{m+1}{p})}{\Gamma(\frac{m+1}{p} + \frac{1}{2})}$$

$$+ \frac{\epsilon\sigma}{2-p} A^{3-p} \left(\frac{1+p}{2p^2} \right)^{\frac{2p}{p+1}} \frac{\Gamma(\frac{1}{p} + \frac{1}{2})}{\Gamma(\frac{1}{p})\Gamma(\frac{1}{2})}$$

$$\cdot \int_{-\infty}^{\infty} \frac{1}{\cosh^{\frac{2}{p}} \tau} \left(\int_{-\infty}^{\tau} \frac{1}{\cosh^{\frac{2}{p}} s} ds \right) d\tau \tag{4.26}$$

$$\frac{dB}{dt} = \frac{2\epsilon\delta p}{2-p} \left(\frac{1+p}{2p^2} \right)^{\frac{m}{p}} \frac{\Gamma(\frac{1}{p} + \frac{1}{2})}{\Gamma(\frac{1}{p})} \frac{\Gamma(\frac{m+1}{p})}{\Gamma(\frac{m+1}{p} + \frac{1}{2})} B^{(\frac{2m+p}{p})}$$

$$+ \frac{\epsilon\sigma p}{2-p} \left(\frac{1+p}{2p^2} \right)^{\frac{1}{p}} \frac{\Gamma(\frac{1}{p} + \frac{1}{2})}{\Gamma(\frac{1}{p})\Gamma(\frac{1}{2})} B^{\frac{2}{p}}$$

$$\cdot \int_{-\infty}^{\infty} \frac{1}{\cosh^{\frac{2}{p}} \tau} \left(\int_{-\infty}^{\tau} \frac{1}{\cosh^{\frac{2}{p}} s} ds \right) d\tau \tag{4.27}$$

$$\frac{d\kappa}{dt} = 0 \tag{4.28}$$

Once again, in (4.26) and (4.27) one can observe that it is necessary to have $p \neq 2$ due to self-focusing singularity.

The velocity of the soliton, as an evolution of the center of mass, is given by

$$v = \frac{d\bar{x}}{dt} = -\kappa + \frac{\epsilon}{E} \int_{-\infty}^{\infty} x(q^*R + q R^*) dx \tag{4.29}$$

which, after integration using (4.8) and (4.20), reduces to

$$v = -\kappa + 2\epsilon\sigma \frac{A^4}{B^3 E} \int_{-\infty}^{\infty} \frac{\tau}{\cosh^{\frac{2}{p}}\tau} \left(\int_{-\infty}^{\tau} \frac{1}{\cosh^{\frac{2}{p}}s} ds \right) d\tau \qquad (4.30)$$

4.4   Quasi-Stationary Solution

In order to obtain the QS solution to (4.19) with R given by (4.20), one can, as in the previous chapter, follow the developments of section 2.5.1. Keeping in mind that $F(s) = s^p$, one gets at the leading order [81, 86]

$$-\left\{ \rho_T^{(0)} + \frac{1}{2}(\rho_X^{(0)})^2 \right\} \hat{q}^{(0)} + \frac{1}{2}\frac{\partial^2 \hat{q}^{(0)}}{\partial \theta^2} + (\hat{q}^{(0)})^{2p+1} = 0 \qquad (4.31)$$

and

$$\left(\rho_X^{(0)} - v^{(0)} \right) \frac{\partial \hat{q}^{(0)}}{\partial \theta} = 0 \qquad (4.32)$$

Now, (4.32) implies

$$\rho_X^{(0)} = v^{(0)} \qquad (4.33)$$

On setting

$$\frac{B^2}{2p^2} = \rho_T^{(0)} + \frac{1}{2}\left(\rho_X^{(0)}\right)^2 = \rho_T^{(0)} + \frac{1}{2}(v^{(0)})^2 \qquad (4.34)$$

(4.31) changes to

$$-\frac{B^2}{2p^2}\hat{q}^{(0)} + \frac{1}{2}\frac{\partial^2 \hat{q}^{(0)}}{\partial \theta^2} + (\hat{q}^{(0)})^{2p+1} = 0 \qquad (4.35)$$

whose solution is

$$\hat{q}^{(0)} = \frac{A}{\cosh^{\frac{1}{p}}[B(\theta - \bar{\theta})]} \qquad (4.36)$$

where

$$B = A^p \left(\frac{2p^2}{1+p} \right)^{\frac{1}{2}} \qquad (4.37)$$

and

$$\frac{d\bar{\theta}}{dt} = v \qquad (4.38)$$

Thus, the soliton frequency and wave number at the first order, as depicted in (4.33) and (4.34), agree with (4.9) and (4.10). At $O(\epsilon)$, decompose

$\hat{q}^{(1)} = \hat{\phi}^{(1)} + i\hat{\psi}^{(1)}$ into its real and imaginary parts. Now the equations for $\hat{\phi}^{(1)}$ and $\hat{\psi}^{(1)}$ are, respectively,

$$-\frac{D^2}{2p^2}\hat{\phi}^{(1)} + \frac{1}{2}\frac{\partial^2 \phi^{(1)}}{\partial\theta^2} + (2p+1)(\hat{q}^{(0)})^{2p}\hat{\phi}^{(1)} = \left(\rho_T^{(1)} + v^{(0)}\rho_X^{(1)}\right)\hat{q}^{(0)} - \frac{\partial^2 \hat{q}^{(0)}}{\partial\theta\partial X}$$

(4.39)

and

$$-\frac{B^2}{2p^2}\hat{\psi}^{(1)} + \frac{1}{2}\frac{\partial^2 \hat{\psi}^{(1)}}{\partial\theta^2} + (\hat{q}^{(0)})^{2p}\hat{\psi}^{(1)}$$

$$= -\frac{\partial\hat{q}^{(0)}}{\partial T} - v^{(0)}\frac{\partial\hat{q}^{(0)}}{\partial X} - \left\{\rho_X^{(1)} - v^{(1)} + \sigma\right\}\frac{\partial\hat{q}^{(0)}}{\partial\theta}$$

$$- \rho_{XX}^{(0)}\hat{q}^{(0)} + \delta(\hat{q}^{(0)})^{2m+1} + \sigma\hat{q}^{(0)}\int_{-\infty}^{x}(\hat{q}^{(0)})^2 ds$$

(4.40)

In (4.40), set $\rho_{XX}^{(0)} = 0$ to eliminate frequency chirp to obtain

$$-\frac{B^2}{2p^2}\hat{\psi}^{(1)} + \frac{1}{2}\frac{\partial^2 \hat{\psi}^{(1)}}{\partial\theta^2} + (\hat{q}^{(0)})^{2p}\hat{\psi}^{(1)}$$

$$= -\frac{\partial\hat{q}^{(0)}}{\partial T} - v^{(0)}\frac{\partial\hat{q}^{(0)}}{\partial X} - \left\{\rho_X^{(1)} - v^{(1)} + \sigma\right\}\frac{\partial\hat{q}^{(0)}}{\partial\theta}$$

$$+ \delta(\hat{q}^{(0)})^{2m+1} + \sigma\hat{q}^{(0)}\int_{-\infty}^{x}(\hat{q}^{(0)})^2 ds$$

(4.41)

Fredholm's alternative, applied to (4.39), yields

$$\frac{\partial B}{\partial X} = 0$$

(4.42)

and

$$\rho_T^{(1)} + v^{(0)}\rho_X^{(1)} = 0$$

(4.43)

whereas, if applied to (4.41), implies

$$\frac{dB}{dT} = \frac{2\delta p}{2-p}\left(\frac{1+p}{2p^2}\right)^{\frac{m}{p}} \frac{\Gamma(\frac{1}{p}+\frac{1}{2})}{\Gamma(\frac{1}{p})} \frac{\Gamma(\frac{m+1}{p})}{\Gamma(\frac{m+1}{p}+\frac{1}{2})} B^{(\frac{2m+p}{p})}$$

$$+ \frac{\sigma p}{2-p}\left(\frac{1+p}{2p^2}\right)^{\frac{1}{p}} \frac{\Gamma(\frac{1}{p}+\frac{1}{2})}{\Gamma(\frac{1}{p})\Gamma(\frac{1}{2})} B^{\frac{2}{p}} \int_{-\infty}^{\infty}\frac{1}{\cosh^{\frac{2}{p}}\tau}\left(\int_{-\infty}^{\tau}\frac{1}{\cosh^{\frac{2}{p}}s}ds\right)d\tau$$

(4.44)

and

$$\rho_X^{(1)} = v^{(1)} - \sigma$$

(4.45)

Equation (4.42) shows that B is a function of T alone and so is A due to (4.37). Thus, by virtue of (4.37), one obtains

$$\frac{dA}{dT} = \frac{2\delta}{2-p} A^{2m+1} \left(\frac{1+p}{2p^2}\right)^{\frac{1}{2p}} \frac{\Gamma(\frac{1}{p}+\frac{1}{2})}{\Gamma(\frac{1}{p})} \frac{\Gamma(\frac{m+1}{p})}{\Gamma(\frac{m+1}{p}+\frac{1}{2})}$$

$$+ \frac{\sigma}{2-p} A^{3-p} \left(\frac{1+p}{2p^2}\right)^{\frac{2p}{p+1}} \frac{\Gamma(\frac{1}{p}+\frac{1}{2})}{\Gamma(\frac{1}{p})\Gamma(\frac{1}{2})} \int_{-\infty}^{\infty} \frac{1}{\cosh^{\frac{2}{p}} \tau} \left(\int_{-\infty}^{\tau} \frac{1}{\cosh^{\frac{2}{p}} s} ds\right) d\tau$$

$$(4.46)$$

Equations (4.44) and (4.46) represent the adiabatic parameter dynamics of the soliton width and amplitude, respectively, in the presence of perturbation terms. Also, note that from these two equations, the perturbation scheme breaks down if $p = 2$. Now (4.39), by virtue of (4.42) and (4.43), reduces to

$$-\frac{B^2}{2p^2} \hat{\phi}^{(1)} + \frac{1}{2} \frac{\partial^2 \hat{\phi}^{(1)}}{\partial \theta^2} + (2p+1)(\hat{q}^{(0)})^{2p} \hat{\phi}^{(1)} = 0 \qquad (4.47)$$

while (4.41), by virtue of (4.42) and (4.45), simplifies to

$$-\frac{B^2}{2p^2} \hat{\psi}^{(1)} + \frac{1}{2} \frac{\partial^2 \hat{\psi}^{(1)}}{\partial \theta^2} + (\hat{q}^{(0)})^{2p} \hat{\psi}^{(1)}$$

$$= -\frac{\partial \hat{q}^{(0)}}{\partial T} + \delta(\hat{q}^{(0)})^{2m+1} + \sigma \hat{q}^{(0)} \int_{-\infty}^{x} (\hat{q}^{(0)})^2 ds \qquad (4.48)$$

Finally, the solutions of (4.42) and (4.43) are, respectively,

$$\hat{\phi}^{(1)} = 0 \qquad (4.49)$$

and

$$\hat{\psi}^{(1)} = -\frac{2B^{\frac{1}{p}}}{p} \left(\frac{1+p}{2p^2}\right)^{\frac{1}{2p}}$$

$$\cdot \left[\frac{\partial \bar{\theta}}{\partial T} \frac{\phi}{\cosh^{\frac{1}{p}} \phi} \int^{\phi} \cosh^{\frac{2}{p}} s_2 \left(\int^{s_2} \frac{\tanh s_1}{\cosh^{\frac{2}{p}} s_1} ds_1 \right) ds_2 \right.$$

$$- \frac{1}{B^2} \frac{dB}{dT} \int^{\phi} \cosh^{\frac{2}{p}} s_2 \left(\int^{s_2} \frac{\tanh s_1}{\cosh^{\frac{2}{p}} s_1} ds_1 \right) ds_2$$

$$\left. + \frac{1}{B^3} \frac{dB}{dT} \int^{\phi} \cosh^{\frac{2}{p}} s_2 \left(\int^{s_2} \frac{\tanh s_1}{\cosh^{\frac{2}{p}} s_1} ds_1 \right) ds_2 \right]$$

$$+ 2\delta B^{\frac{2m-2p+1}{p}} \left(\frac{1+p}{2p^2}\right)^{\frac{2m+1}{2p}} \int^{\phi} \cosh^{\frac{2}{p}} s_2 \left(\int^{s_2} \frac{\tanh s_1}{\cosh^{(\frac{2m+2}{p})} s_1} ds_1 \right) ds_2$$

$$+ 2\sigma \frac{A^3}{B^3} \frac{1}{\cosh^{\frac{1}{p}} \phi} \int^{\phi} \cosh^{\frac{2}{p}} s_3$$

$$\cdot \left(\int^{s_3} \frac{1}{\cosh^{\frac{2}{p}} s_2} \left(\int_{-\infty}^{s_2} \frac{1}{\cosh^{\frac{2}{p}} s_1} ds_1 \right) ds_2 \right) ds_3 \tag{4.50}$$

which leads to the QS solution given by (2.98).

The fixed point of the dynamical system given by (4.26) and (4.28) (or by (4.27) and (4.28)) is

$$\bar{A} = \left[\frac{\sigma I_p}{2\delta} \left(\frac{1+p}{2p^2} \right)^{\frac{1}{2}} \frac{\Gamma\left(\frac{m+1}{p} + \frac{1}{2} \right)}{\Gamma\left(\frac{m+1}{p} \right) \Gamma\left(\frac{1}{2} \right)} \right]^{\frac{1}{2m+p-2}} \tag{4.51}$$

and

$$\bar{B} = \left[\frac{\sigma I_p}{2\delta} \left(\frac{2p^2}{1+p} \right)^{\frac{m-1}{p}} \frac{\Gamma\left(\frac{m+1}{p} + \frac{1}{2} \right)}{\Gamma\left(\frac{m+1}{p} \right) \Gamma\left(\frac{1}{2} \right)} \right]^{\frac{p}{2m+p-2}} \tag{4.52}$$

with

$$I_p = \int_{-\infty}^{\infty} \frac{1}{\cosh^{\frac{2}{p}} \tau} \left(\int_{-\infty}^{\tau} \frac{1}{\cosh^{\frac{2}{p}} s} ds \right) d\tau \tag{4.53}$$

This fixed point is a *sink*. This physically means that the QS soliton for power law nonlinearity travels through an optical fiber with the velocity given by (4.30) at a fixed amplitude and width as in (4.51) and (4.52).

Exercises

1. Consider the following NLSE, with filters as perturbation terms

$$iq_t + \frac{1}{2}q_{xx} + |q|^{2p}q = i\epsilon\beta q_{xx}$$

Prove that the adiabatic variation of the amplitude and frequency are respectively given by

$$\frac{dA}{dt} = \frac{2\epsilon\beta}{2-p} \frac{A^p}{B} \frac{1}{} \left(\frac{2p^2}{p+1} \right)^{\frac{p-1}{2p}} \frac{\Gamma\left(\frac{1}{p} + \frac{1}{2} \right)}{\Gamma\left(\frac{1}{p} \right)}$$

$$\cdot \left[\frac{A^2 B^2}{p^2} \frac{\Gamma\left(\frac{p+1}{p} \right)}{\Gamma\left(\frac{p+1}{p} + \frac{1}{2} \right)} - \frac{A^2}{p^2} (\kappa^2 p^2 + B^2) \frac{\Gamma\left(\frac{1}{p} \right)}{\Gamma\left(\frac{1}{p} + \frac{1}{2} \right)} \right]$$

and

$$\frac{d\kappa}{dt} = \frac{4\epsilon\beta}{p^2} \kappa A^2 B^{\frac{2p-2}{p}} \left(\frac{2p^2}{p+1} \right)^{\frac{1}{2}} \left(\frac{p-2}{p+2} \right)$$

2. For the power law nonlinearity with Hamiltonian perturbations given by

$$iq_t + \frac{1}{2}q_{xx} + |q|^{2p}q = i\epsilon[\lambda(|q|^2 q)_x + v(|q|^2)_x q - \gamma q_{xxx}]$$

prove that the velocity of the soliton is given by

$$v = -\kappa - \frac{\epsilon}{4}(3\lambda + 2v)\frac{\Gamma(\frac{1}{2} + \frac{1}{p})\,\Gamma(\frac{2}{p})}{\Gamma(\frac{1}{2} + \frac{2}{p})\,\Gamma(\frac{1}{p})}$$

$$-\frac{3\epsilon\gamma}{4p^2} + \frac{3\epsilon\gamma}{p^2}\frac{\Gamma(\frac{p-1}{p})\,\Gamma(\frac{1}{2} + \frac{1}{p})}{\Gamma(\frac{1}{p})\,\Gamma(\frac{3}{2} + \frac{1}{p})}$$

3. Prove that the fixed point of the dynamical system given by (4.26) and (4.28) is given by $(\bar{A}, 0)$ where

$$\bar{A} = \left[\frac{\sigma I_p}{2\delta}\left(\frac{1+p}{2p^2}\right)^{\frac{1}{2}}\frac{\Gamma(\frac{m+1}{p} + \frac{1}{2})}{\Gamma(\frac{m+1}{p})\Gamma(\frac{1}{2})}\right]^{\frac{1}{2m+p-2}}$$

4. Consider the moment of inertia of the soliton that is given by

$$J = \int_{-\infty}^{\infty} x^2 |q|^2 dx$$

Prove that J satisfies the relation

$$\frac{d^2 J}{dt^2} = 8H$$

where H is the Hamiltonian.

5

Parabolic Law Nonlinearity

This chapter talks about the optical solitons of the nonlinear Schrödinger's equation (NLSE) with parabolic law nonlinearity, commonly known as *cubic-quintic nonlinearity*. Section 5.1 contains a detailed discussion of the physics of parabolic law and the mathematical aspects of the equation. Section 5.2 reviews the three conserved quantities of the NLSE with parabolic law nonlinearity. Also discussed here are the adiabatic dynamics of the optical soliton parameters due to perturbation terms. Finally, Section 5.3 discusses the quasi-stationary solution of the perturbed NLSE with parabolic law.

5.1 Introduction

In this chapter, the NLSE with parabolic law nonlinearity will be studied. It has been known for a long time that optical beams can self-focus in both space and time while propagating in a nonlinear medium. The collapse of two- and three-dimensional optical beams in a Kerr law medium was considered as a means of producing high electric field strengths. It was observed that the inclusion of a saturable nonlinearity could halt the singular collapse, thus causing the formation of an optical beam that propagates without changing its temporal or spatial shape, held together by nonlinear effects.

To obtain some knowledge of the diameter of the self-trapping beam, it is necessary to consider nonlinearities higher than the third order. It was recognized in 1960s and 1970s that saturation of the nonlinear refractive index plays a fundamental role in the self-trapping phenomenon. Higher order nonlinearities arise by retaining the higher order terms in the nonlinear polarization tensor. For a fifth order nonlinearity, the refractive index is given by [82]

$$n = n_0 + n_2|E|^2 + n_4|E|^4 \tag{5.1}$$

where n_0 is the linear refractive index of the medium and $|E|^2$ is the electric field intensity of the light wave, while $n_2 = 3\chi^{(3)}/8n_0$ and $n_4 = 5\chi^{(5)}/16n_0$ with $n_0 > n_2|E|^2 > n_4|E|^4$. Here, n_2 and n_4 respectively represent the third- and fifth-order nonlinear coefficients. In general, the coefficients n_2 and n_4 could be positive or negative, depending on the medium and the frequency selected.

Little attention was paid to the propagation of optical beams in fifth-order nonlinear media because no analytic solutions were known and it seemed that the chances of finding any material with a significant fifth-order term were slim. However, recent developments have rekindled interest in this area. The optical susceptibility of CdS_xSe_{1-x}-doped glasses was experimentally shown to have a considerable $\chi^{(5)}$, or fifth-order susceptibility. It was also demonstrated that a significant $\chi^{(5)}$ nonlinearity effect exists in a transparent glass in intense femtosecond pulses at 620 nm.

Besides these saturation effects, it was also proposed that the doping of silica glass with two appropriate semiconductor particles leads (in a region far from saturation) exactly to the parabolic form of the refractive index with an effective increased value of n_4 and a reduced value of n_2. In other materials, the values of n_2 and n_4 may also depend on the doping.

The organic nonlinear material polydiacetylene para-toluene sulfonate (PTS) is another material that can be described exactly by the parabolic law. Measurements of PTS indicate that $n_2 > 0$ and $n_4 > 0$ at 1600 nm, with low loss and negligible multisoliton absorption. The spectrum of PTS reveals three two-photon excited states that can account for the observed value of n_2. The absence of nearby states that can saturate can be interpreted as n_4 arising from the Stark shift of the two-photon states. This leads to the conclusion that the parabolic law model can be considered as an exact model for PTS and similar materials and this material yields solitons at lower power than has previously been achieved. The experimental confirmation places PTS in the unique class of solid-state materials with positive third-order and negative fifth-order nonlinearity.

A recent study of the nonlinear self-phase modulation of a fundamental beam involved in a process of third-harmonic generation showed that the beam experienced an additional higher order phase shift as a consequence of the n_4 coefficient being a sum of two terms. The first is due to the inherent $\chi^{(5)}$ susceptibility of the medium and the second is due to the cascading of the third-order nonlinearity, namely

$$n_4 = n_4^{dir} + n_4^{casc}$$

with the cascading term n_4^{casc} being proportional to $(\chi^{(3)})^2$. The cascading term n_4^{casc} can always be made to exceed n_4^{dir} by manipulating the sample length, wave-vector mismatch, and beam intensity.

These developments and sustained theoretical interest in optical beam propagation motivated the study of solitons in a parabolic medium. The dimensionless form of the NLSE with parabolic law nonlinearity is given by [82, 86]

$$iq_t + \frac{1}{2}q_{xx} + (|q|^2 + \nu|q|^4)q = 0 \tag{5.2}$$

where v is a constant. For $v = 0$, the Kerr law of nonlinearity is recovered. Equation (5.2) is not integrable by the inverse scattering transform. However, one can solve it by the traveling wave ansatz.

5.2 Traveling Wave Solution

For the case of parabolic law nonlinearity, equation (2.1) reduces to

$$iq_t + \frac{1}{2}q_{xx} + (|q|^2 + v|q|^4)q = 0 \tag{5.3}$$

so that equation (2.10) simplifies to

$$B^2 g'' - (\kappa^2 - 2\omega)g + 2A^2 g^3 + 2v A^4 g^5 = 0 \tag{5.4}$$

Multiplying (5.4) by g' and integrating yields

$$(g')^2 = \frac{g^2}{3B^2}[3(\kappa^2 - 2\omega) - 3A^3 g^2 - 2v A^4 g^4] \tag{5.5}$$

Separating variables and integrating yields

$$x - vt = \sqrt{3} \int \frac{dg}{g[3(\kappa^2 - 2\omega) - 3A^3 g^2 - 2v A^4 g^4]^{\frac{1}{2}}} \tag{5.6}$$

which leads to the 1-soliton solution

$$q(x, t) = \frac{A}{[1 + a \cosh\{B(x - \bar{x}(t))\}]^{\frac{1}{2}}} e^{i(-\kappa x + \omega t + \sigma_0)} \tag{5.7}$$

where

$$B(t) = \sqrt{2}A(t) \tag{5.8}$$

$$\kappa = -v \tag{5.9}$$

$$\omega = \frac{A^2}{4} - \frac{\kappa^2}{2} \tag{5.10}$$

and

$$a = \sqrt{1 + \frac{4}{3}v A^2} \tag{5.11}$$

Here, A is the amplitude of the soliton, B is its width, v is its velocity, κ is the soliton frequency, ω is the wave number, and $\bar{x}$ and σ_0 are the center of the soliton and the center of the soliton phase, respectively. Equation (5.8) gives the relation between the amplitude and width of the soliton for the case of power law nonlinearity. Thus, the soliton parameter dynamics are

$$\frac{dA}{dt} = 0 \tag{5.12}$$

$$\frac{dB}{dt} = 0 \tag{5.13}$$

$$\frac{d\kappa}{dt} = 0 \tag{5.14}$$

and

$$\frac{d\bar{x}}{dt} = -\kappa \tag{5.15}$$

5.3   Integrals of Motion

The NLSE with parabolic law nonlinearity has three integrals of motion. These are obtained from (2.41)–(2.43) by setting $F(s) = s + \nu s^2$ so that $f(s) = s^2/2 + \nu s^3/3$. Thus the three integrals of motion are

$$E = \int_{-\infty}^{\infty} |q|^2 dx = \begin{cases} \frac{A^2}{B}\sqrt{\frac{3}{2\nu}}\tan^{-1}\left[A\sqrt{\frac{\nu\omega}{3}}\right] : 0 < \nu < \infty \\ \frac{A^2}{B}\sqrt{-\frac{3}{2\nu}}\tanh^{-1}\left[A\sqrt{-\frac{\nu\omega}{3}}\right] : -\frac{3}{16\omega} < \nu < 0 \end{cases} \tag{5.16}$$

$$M = \frac{i}{2}\int_{-\infty}^{\infty}(qq_x^* - q^*q_x)dx$$
$$= \begin{cases} -\frac{\kappa}{2}\frac{A^2}{B}\sqrt{\frac{3}{2\nu}}\tan^{-1}\left[A\sqrt{\frac{\nu\omega}{3}}\right] : 0 < \nu < \infty \\ -\frac{\kappa}{2}\frac{A^2}{B}\sqrt{-\frac{3}{2\nu}}\tanh^{-1}\left[A\sqrt{-\frac{\nu\omega}{3}}\right] : -\frac{3}{16\omega} < \nu < 0 \end{cases} \tag{5.17}$$

and

$$H = \int_{-\infty}^{\infty}\left[\frac{1}{2}|q_x|^2 - \frac{1}{2}|q|^4 - \frac{\nu}{3}|q|^6\right]dx$$
$$= \begin{cases} -\frac{3\sqrt{2\omega}}{8\nu} + \frac{3}{8\nu}\sqrt{\frac{3}{2\nu}}\tan^{-1}\left[\frac{-\sqrt{3}+\sqrt{3+16\nu\omega}}{4\sqrt{\nu\omega}}\right] : 0 < \nu < \infty \\ -\frac{3\sqrt{2\omega}}{8\nu} - \frac{3}{8\nu}\sqrt{-\frac{3}{2\nu}}\tanh^{-1}\left[\frac{-\sqrt{3}+\sqrt{3-16\nu\omega}}{4\sqrt{-\nu\omega}}\right] : -\frac{3}{16\omega} < \nu < 0 \end{cases} \tag{5.18}$$

In this chapter, the perturbed NLSE with parabolic law nonlinearity that is going to be studied is

$$iq_t + \frac{1}{2}q_{xx} + (|q|^2 + v|q|^4)q = i\epsilon R[q, q^*] \tag{5.19}$$

where

$$R = \delta|q|^{2m}q + \sigma q \int_{-\infty}^{x} |q|^2 ds \tag{5.20}$$

In the presence of the perturbation term in (5.20), the soliton parameters deform adiabatically. These adiabatic variations are given by

$$\frac{dA}{dt} = \epsilon a^2 \frac{\sqrt{2}}{2} \int_{-\infty}^{\infty} (q^*R + q R^*)dx \tag{5.21}$$

$$\frac{dB}{dt} = \epsilon a^2 \int_{-\infty}^{\infty} (q^*R + q R^*)dx \tag{5.22}$$

and

$$\frac{d\kappa}{dt} = \epsilon \frac{B}{A^2 E} \left[i \int_{-\infty}^{\infty} (q_x^*R - q_x R^*)dx - \kappa \int_{-\infty}^{\infty} (q^*R + q R^*)dx \right] \tag{5.23}$$

where E is the energy of the soliton given by (5.16). From (2.51) and (2.52), the adiabatic variations of the energy and the linear momentum of the soliton due to parabolic law nonlinearity take the forms

$$\frac{dE}{dt} = \epsilon \left[\frac{\delta\sqrt{2}A^{2m+1}}{2^m a^{m+1}} F\left(m+1, m+1, m+\frac{3}{2}; \frac{a-1}{2a}\right) B\left(m+1, \frac{1}{2}\right) \right.$$
$$\left. + \frac{2\sigma A^4}{B^2} \int_{-\infty}^{\infty} \frac{1}{1 + a\cosh\tau} \left(\int_{-\infty}^{\tau} \frac{1}{1 + a\cosh s} ds \right) d\tau \right] \tag{5.24}$$

and

$$\frac{dM}{dt} = \epsilon\kappa \left[\frac{\delta\sqrt{2}A^{2m+1}}{2^m a^{m+1}} F\left(m+1, m+1, m+\frac{3}{2}; \frac{a-1}{2a}\right) B\left(m+1, \frac{1}{2}\right) \right.$$
$$\left. + \frac{2\sigma A^4}{B^2} \int_{-\infty}^{\infty} \frac{1}{1 + a\cosh\tau} \left(\int_{-\infty}^{\tau} \frac{1}{1 + a\cosh s} ds \right) d\tau \right] \tag{5.25}$$

where $F(\alpha, \beta; \gamma; z)$ is the Gauss' hypergeometric function defined as

$$F(\alpha, \beta; \gamma; z) = \frac{\Gamma(\gamma)}{\Gamma(\alpha)\Gamma(\beta)} \sum_{n=0}^{\infty} \frac{\Gamma(\alpha+n)\Gamma(\beta+n)}{\Gamma(\gamma+n)} \frac{z^n}{n!} \tag{5.26}$$

and $B(l, m)$ is the usual beta function. Substituting the perturbation term R from (5.20) and carrying out the integrations in (5.21)–(5.23) yields

$$\frac{dA}{dt} = \frac{\epsilon \delta A^{2m+1}}{2^m a^{m-1}} F\left(m+1, m+1, m+\frac{3}{2}; \frac{a-1}{2a}\right) B\left(m+1, \frac{1}{2}\right)$$

$$+ \epsilon \sigma a^2 \frac{\sqrt{2}}{2} \int_{-\infty}^{\infty} \frac{1}{1+a\cosh\tau} \left(\int_{-\infty}^{\tau} \frac{1}{1+a\cosh s} ds\right) d\tau \qquad (5.27)$$

$$\frac{dB}{dt} = \frac{\epsilon \delta \sqrt{2} A^{2m+1}}{2^m a^{m-1}} F\left(m+1, m+1, m+\frac{3}{2}; \frac{a-1}{2a}\right) B\left(m+1, \frac{1}{2}\right)$$

$$+ \epsilon \sigma a^2 \int_{-\infty}^{\infty} \frac{1}{1+a\cosh\tau} \left(\int_{-\infty}^{\tau} \frac{1}{1+a\cosh s} ds\right) d\tau \qquad (5.28)$$

and

$$\frac{d\kappa}{dt} = 0 \qquad (5.29)$$

The velocity of the soliton, as the evolution of the center of mass, is given by

$$v = \frac{d\bar{x}}{dt} = -\kappa + \frac{\epsilon}{E} \int_{-\infty}^{\infty} x(q^* R + q R^*) dx \qquad (5.30)$$

which, after integration, or by directly using (2.67), reduces to

$$v = -\kappa + 2\epsilon \sigma \frac{A^4}{BE} \int_{-\infty}^{\infty} \frac{\tau}{1+a\cosh\tau} \left(\int_{-\infty}^{\tau} \frac{1}{1+a\cosh s} ds\right) d\tau \qquad (5.31)$$

5.4 Quasi-Stationary Solution

In this section, the QS solution to the perturbed NLSE with parabolic law nonlinearity will be obtained. Following the method that was developed in section 2.5, one gets at the leading order

$$-\left\{\rho_T^{(0)} + \frac{1}{2}\left(\rho_X^{(0)}\right)^2\right\} \hat{q}^{(0)} + \frac{1}{2}\frac{\partial^2 \hat{q}^{(0)}}{\partial \theta^2} + (\hat{q}^{(0)})^3 + \nu(\hat{q}^{(0)})^5 = 0 \qquad (5.32)$$

and

$$\left(\rho_X^{(0)} - v^{(0)}\right) \frac{\partial \hat{q}^{(0)}}{\partial \theta} = 0 \qquad (5.33)$$

Now, (5.33) implies

$$\rho_X^{(0)} = v^{(0)} \qquad (5.34)$$

On setting

$$\frac{B^2}{8} = \rho_T^{(0)} + \frac{1}{2}\left(\rho_X^{(0)}\right)^2 = \rho_T^{(0)} + \frac{1}{2}(v^{(0)})^2 \qquad (5.35)$$

(5.32) changes to

$$-\frac{B^2}{8}\hat{q}^{(0)} + \frac{1}{2}\frac{\partial^2\hat{q}^{(0)}}{\partial\theta^2} + (\hat{q}^{(0)})^3 + v(\hat{q}^{(0)})^5 = 0 \tag{5.36}$$

whose solution is

$$\hat{q}^{(0)} = Ag[B(\theta - \bar{\theta})] \tag{5.37}$$

where

$$g(\tau) = \frac{1}{\left[1 + \sqrt{1 + \frac{4}{3}vA^2}\cosh\tau\right]^{\frac{1}{2}}} \tag{5.38}$$

$$B = A\sqrt{2} \tag{5.39}$$

$$\tau = B(\theta - \bar{\theta}) \tag{5.40}$$

and

$$\frac{d\bar{\theta}}{dt} = v \tag{5.41}$$

At $O(\epsilon)$ level, decomposing $\hat{q}^{(1)} = \hat{\phi}^{(1)} + i\hat{\psi}^{(1)}$ into its real and imaginary parts, the equations for $\hat{\phi}^{(1)}$ and $\hat{\psi}^{(1)}$, by virtue of (5.36), are, respectively,

$$-\frac{B^2}{8}\hat{\phi}^{(1)} + \frac{1}{2}\frac{\partial^2\hat{\phi}^{(1)}}{\partial\theta^2} + \{3(\hat{q}^{(0)})^2 + 5v(\hat{q}^{(0)})^4\}\hat{\phi}^{(1)}$$

$$= \{\rho_T^{(1)} + v^{(0)}\rho_X^{(1)}\}\hat{q}^{(0)} - \frac{\partial^2\hat{q}^{(0)}}{\partial\theta\partial X} \tag{5.42}$$

and

$$-\frac{B^2}{8}\hat{\psi}^{(1)} + \frac{1}{2}\frac{\partial^2\hat{\psi}^{(1)}}{\partial\theta^2} + \{(\hat{q}^{(0)})^2 + v(\hat{q}^{(0)})^4\}\hat{\psi}^{(1)} = -\frac{\partial\hat{q}^{(0)}}{\partial T} - v^{(0)}\frac{\partial\hat{q}^{(0)}}{\partial X}$$

$$-\{\rho_X^{(1)} - v^{(1)} + \sigma\}\frac{\partial\hat{q}^{(0)}}{\partial\theta} - \rho_{XX}^{(0)}\hat{q}^{(0)} - \delta(\hat{q}^{(0)})^{2m+1}$$

$$+\sigma\hat{q}^{(0)}\int_{-\infty}^{x}(\hat{q}^{(0)})^2ds \tag{5.43}$$

In (5.43), set $\rho_{XX}^{(0)} = 0$ to eliminate frequency chirp to obtain

$$-\frac{B^2}{8}\hat{\psi}^{(1)} + \frac{1}{2}\frac{\partial^2\hat{\psi}^{(1)}}{\partial\theta^2} + \{(\hat{q}^{(0)})^2 + v(\hat{q}^{(0)})^4\}\hat{\psi}^{(1)} = -\frac{\partial\hat{q}^{(0)}}{\partial T} - v^{(0)}\frac{\partial\hat{q}^{(0)}}{\partial X}$$

$$-\{\rho_X^{(1)} - v^{(1)} + \sigma\}\frac{\partial\hat{q}^{(0)}}{\partial\theta} - \delta(\hat{q}^{(0)})^{2m+1} + \sigma\hat{q}^{(0)}\int_{-\infty}^{x}(\hat{q}^{(0)})^2ds \tag{5.44}$$

Fredholm's alternative, when applied to (5.42) yields

$$\frac{\partial B}{\partial X} = 0 \tag{5.45}$$

and

$$\rho_T^{(1)} + v^{(0)} \rho_X^{(1)} = 0 \tag{5.46}$$

whereas if applied to (5.44) implies

$$\frac{dB}{dT} = \frac{\delta \sqrt{2} A^{2m+1}}{2^m a^{m-1}} F\left(m+1, m+1, m+\frac{3}{2}; \frac{a-1}{2a}\right) B\left(m+1, \frac{1}{2}\right)$$

$$+ \sigma a^2 \int_{-\infty}^{\infty} \frac{1}{1 + a \cosh \tau} \left(\int_{-\infty}^{\tau} \frac{1}{1 + a \cosh s} ds \right) d\tau \tag{5.47}$$

and

$$\rho_X^{(1)} = v^{(1)} - \sigma \tag{5.48}$$

Equation (5.45) shows that B is a function of T only, and so is A by virtue of (5.39) so that

$$\frac{dA}{dT} = \frac{\delta A^{2m+1}}{2^m a^{m-1}} F\left(m+1, m+1, m+\frac{3}{2}; \frac{a-1}{2a}\right) B\left(m+1, \frac{1}{2}\right)$$

$$+ \sigma a^2 \frac{\sqrt{2}}{2} \int_{-\infty}^{\infty} \frac{1}{1 + a \cosh \tau} \left(\int_{-\infty}^{\tau} \frac{1}{1 + a \cosh s} ds \right) d\tau \tag{5.49}$$

Again, note that (5.49) and (5.47) are the same as (5.27) and (5.28), respectively, while relations (5.46) and (5.48) cannot be recovered by the soliton perturbation theory. The $O(\epsilon)$ equations now reduce to

$$-\frac{B^2}{8} \hat{\phi}^{(1)} + \frac{1}{2} \frac{\partial^2 \hat{\phi}^{(1)}}{\partial \theta^2} + \{3(\hat{q}^{(0)})^2 + 5v(\hat{q}^{(0)})^4\} \hat{\phi}^{(1)} = 0 \tag{5.50}$$

and

$$-\frac{B^2}{8} \hat{\psi}^{(1)} + \frac{1}{2} \frac{\partial^2 \hat{\psi}^{(1)}}{\partial \theta^2} + \{(\hat{q}^{(0)})^2 + v(\hat{q}^{(0)})^4\} \hat{\psi}^{(1)}$$

$$= -\frac{\partial \hat{q}^{(0)}}{\partial T} - \delta(\hat{q}^{(0)})^{2m+1} + \sigma \hat{q}^{(0)} \int_{-\infty}^{x} (\hat{q}^{(0)})^2 \, ds \tag{5.51}$$

whose solutions are, respectively,

$$\hat{\phi}^{(1)} = 0 \tag{5.52}$$

and

$$\hat{\psi}^{(1)} = -\frac{1}{\sqrt{2}}\frac{\partial\bar{\theta}}{\partial T}\frac{1}{(1+a\cosh\tau)^{\frac{1}{2}}}\int^{\tau}(1+a\cosh s_2)\left(\int^{s_2}\frac{a\sinh s_1}{1+a\cosh s_1}ds_1\right)ds_2$$

$$+\frac{1}{2A^2}\frac{dA}{dT}\frac{1}{(1+a\cosh\tau)^{\frac{1}{2}}}\int^{\tau}(1+a\cosh s_2)$$

$$\left(\int^{s_2}\frac{as_1\sinh s_1}{(1+a\cosh s_1)^2}ds_1\right)ds_2 - \frac{1}{2A^2}\frac{dA}{dT}\frac{1}{(1+a\cosh\tau)^{\frac{1}{2}}}$$

$$\cdot\int^{\tau}(1+a\cosh s_2)\left(\int^{s_2}\frac{1}{1+a\cosh s_1}ds_1\right)ds_2 - \delta A^{2m-1}\frac{1}{(1+a\cosh\tau)^{\frac{1}{2}}}$$

$$\cdot\int^{\tau}(1+a\cosh s_2)\left(\int^{s_2}\frac{as_1\sinh s_1}{(1+a\cosh s_1)^{m+1}}ds_1\right)ds_2$$

$$+2\sigma\frac{A^3}{B^3}\frac{1}{(1+a\cosh\tau)^{\frac{1}{2}}}\int^{\phi}(1+a\cosh s_3)$$

$$\cdot\left(\int^{s_3}\frac{1}{1+a\cosh s_2}\left(\int_{-\infty}^{s_2}\frac{1}{1+a\cosh s_1}ds_1\right)ds_2\right)ds_3 \qquad (5.53)$$

which lead to the QS soliton that is given by (2.98).

The dynamical system in (5.27) and (5.29) or (5.28) and (5.29) has a stable fixed point, so that the QS soliton for parabolic law nonlinearity propagates through an optical fiber with a velocity given by (5.31).

Exercises

1. Prove that as v approaches zero, the energy of the parabolic law soliton approaches the energy of the Kerr law soliton—in other words,

$$\lim_{v\to0} E = 2A$$

Remember to consider the right-hand and left-hand limits separately.

2. Prove that the parabolic law soliton that is given by (5.7) is equivalent to the Kerr law soliton with $v = 0$.

3. Prove that for the parabolic law NLSE with filters as perturbation terms

$$iq_t + \frac{1}{2}q_{xx} + (|q|^2 + v|q|^4)q = i\epsilon\beta q_{xx}$$

the adiabatic parameter dynamics of the soliton amplitude and frequency are given by

$$\frac{dA}{dt} = -4\sqrt{2}\epsilon\beta\frac{a^2 B^2}{\sqrt{a^2-1}}\tan^{-1}\sqrt{\frac{a-1}{a+1}}$$

$$-\epsilon\beta a^2 A^3\left[\frac{a^2}{(a^2-1)^{\frac{3}{2}}}\tan^{-1}\sqrt{\frac{a-1}{a+1}} - \frac{1}{2(a^2-1)}\right]$$

and

$$\frac{d\kappa}{dt} = -\epsilon\beta\frac{\sqrt{2}A^3 B}{E}\left[\frac{a^2}{(a^2-1)^{\frac{3}{2}}}\tan^{-1}\sqrt{\frac{a-1}{a+1}} - \frac{1}{2(a^2-1)}\right]$$

where E is the energy of the soliton.

4. Consider the Hamiltonian-type perturbations for the NLSE with parabolic law nonlinearity

$$iq_t + \frac{1}{2}q_{xx} + (|q|^2 + v|q|^4)q = i\epsilon[\lambda(|q|^2 q)_x + \mu(|q|^2)_x q - \gamma q_{xxx}]$$

Prove that the velocity is given by

$$v = -\kappa - \frac{\epsilon\sqrt{2}}{2E}\frac{A^3}{(a^2-1)^{\frac{3}{2}}}(3\lambda+2\mu)\left[\sqrt{a^2-1} - 2\tan^{-1}\sqrt{\frac{a-1}{a+1}}\right]$$

$$-3\sqrt{2}\frac{\epsilon\gamma A}{E}\left[\left(\frac{a^4 A^2}{2} + \frac{2\kappa^2}{\sqrt{a^2-1}}\right)\tan^{-1}\sqrt{\frac{a-1}{a+1}} - \frac{a^2 A^2}{4}\sqrt{a^2-1}\right]$$

5. Prove that the series for Gauss' hypergeometric function $F(a, b; c; x)$ that is defined in (5.26) is convergent for $c > a + b$.

6

Dual-Power Law Nonlinearity

This chapter talks about the optical solitons of the nonlinear Schrödinger's equation (NLSE) with dual-power law nonlinearity, a generalization of the parabolic law of nonlinearity. Section 6.1 contains a brief discussion of the physics of the dual-power law and the mathematical results of the optical solitons with dual-power law. Section 6.2 talks about the three conserved quantities of the NLSE with dual-power law. Also in this section are the results of the adiabatic parameter dynamics of the solitons in presence of perturbation terms. Finally, Section 6.3 talks about the quasi-stationary solution (QS) of the perturbed NLSE.

6.1 Introduction

The NLSE with the dual-power law of nonlinearity appears in various areas of mathematical physics and nonlinear optics. This model is used to describe the saturation of the nonlinear refractive index. It also serves as a basic model to describe the solitons in photovoltaic-photorefractive materials such as lithium niobate. The propagation of ultrashort optical pulses in a nonlinear medium can be characterized by the nonlinear refractive index that is given by [88]

$$n = n_0 + n_2|E|^{2p} + n_4|E|^{4p} \qquad (6.1)$$

Here, also as in the parabolic law case, $n_0 > n_2|E|^{2p} > n_4|E|^{4p}$. The dimensionless form of the NLSE with dual-power law of nonlinearity is

$$iq_t + \frac{1}{2}q_{xx} + (|q|^{2p} + v|q|^{4p})q = 0 \qquad (6.2)$$

It should be noted that in (6.2), $v = 0$ reduces to the case of power law nonlinearity, and if in addition $p = 1$, the case of Kerr law nonlinearity is recovered. If, however, $v \neq 0$ and $p = 1$, one falls back to the case of parabolic law nonlinearity that was studied in the previous chapter. Thus, the case of dual-power law is the most generalized case whose exact soliton solution is known that is studied here. In the next chapter, a different kind of nonlinearity will be studied whose exact soliton solution is not known. Although the dual-power law and parabolic law have been further extended to higher order polynomial law and triple-power law nonlinearity, their exact soliton solutions are not yet known. Studies are under way to obtain an exact soliton solution for these laws of nonlinearity.

Although (6.2) is not integrable by the inverse scattering transform, one can apply the traveling wave technique to obtain the 1-soliton solution of the NLSE with dual-power law nonlinearity.

6.2 Traveling Wave Solution

For the case of dual-power law nonlinearity, (2.1) reduces to

$$iq_t + \frac{1}{2}q_{xx} + (|q|^{2p} + |q|^{4p})q = 0 \tag{6.3}$$

so that (2.10) simplifies to

$$B^2 g'' - (\kappa^2 - 2\omega)g + 2A^{2p}g^{2p+1} + 2\nu A^{4p}g^{4p+1} = 0 \tag{6.4}$$

Multiplying (6.4) by g' and integrating yields

$$(g')^2 = \frac{g^2}{B^2(p+1)(2p+1)}[(p+1)(2p+1)(\kappa^2 - 2\omega) - 2(2p+1)A^{2p}g^{2p}$$
$$- 2\nu(p+1)A^{4p}g^{4p}] \tag{6.5}$$

Separating variables and integrating yields

$$x - vt$$
$$= \int \frac{\sqrt{(p+1)(2p+1)}dg}{[(p+1)(2p+1)(\kappa^2 - 2\omega) - 2(2p+1)A^{2p}g^{2p} - 2\nu(p+1)A^{4p}g^{4p}]^{\frac{1}{2}}} \tag{6.6}$$

which leads to the 1-soliton solution

$$q(x,t) = \frac{A}{[1 + a\cosh\{B(x - \bar{x}(t))\}]^{\frac{1}{2p}}}e^{i\{-\kappa x + \omega t + \sigma_0\}} \tag{6.7}$$

where

$$B = A^p\left(\frac{4p^2}{1+p}\right)^{\frac{1}{2p}} \tag{6.8}$$

with

$$\kappa = -v \tag{6.9}$$

$$\omega = \frac{A^{2p}}{2p+2} - \frac{\kappa^2}{2} \tag{6.10}$$

and

$$a = \sqrt{1 + \frac{vB^2}{2p^2} \frac{(1+p)^2}{1+2p}} \tag{6.11}$$

Here, A is the amplitude of the soliton, B is its width, v is its velocity, κ is the soliton frequency, ω is the wave number, and $\bar{x}$ and σ_0 are the center of the soliton and the center of the soliton phase, respectively. For dual-power law nonlinearity, solitons exist for

$$-\frac{2p^2}{B^2} \frac{1+2p}{(1+p)^2} < v < 0 \tag{6.12}$$

The corresponding parameter dynamics for the solitons are given by

$$\frac{dA}{dt} = 0 \tag{6.13}$$

$$\frac{dB}{dt} = 0 \tag{6.14}$$

$$\frac{d\kappa}{dt} = 0 \tag{6.15}$$

and

$$\frac{d\bar{x}}{dt} = -\kappa \tag{6.16}$$

6.3 Integrals of Motion

For dual-power law nonlinearity, setting $F(s) = s^p + vs^{2p}$, so that $f(s) = s^{p+1}/p+1+vs^{2p+1}/2p+1$, (2.41)–(2.43) gives the three conserved quantities as

$$E = \int_{-\infty}^{\infty} |q|^2 dx = \frac{2A^2}{B2^{\frac{1}{p}}a^{\frac{1}{p}}} F\left(\frac{1}{p}, \frac{1}{p}; \frac{1}{2} + \frac{1}{p}; \frac{a-1}{2a}\right) B\left(\frac{1}{p}, \frac{1}{2}\right) \tag{6.17}$$

$$M = \frac{i}{2} \int_{-\infty}^{\infty} (qq_x^* - q^*q_x) dx$$

$$= -\frac{2\kappa A^2}{B2^{\frac{1}{p}}a^{\frac{1}{p}}} F\left(\frac{1}{p}, \frac{1}{p}; \frac{1}{2} + \frac{1}{p}; \frac{a-1}{2a}\right) B\left(\frac{1}{p}, \frac{1}{2}\right) \tag{6.18}$$

and

$$H = \int_{-\infty}^{\infty} \left[\frac{1}{2}|q_x|^2 - \frac{|q|^{2p+2}}{p+1} - v\frac{|q|^{4p+2}}{2p+1} \right] dx$$

$$= \frac{A^2}{2^{\frac{1}{p}} a^{\frac{1}{p}}} \left[\frac{B}{4p^2} F\left(2 + \frac{1}{p}, \frac{1}{p}; \frac{3}{2} + \frac{1}{p}; \frac{a-1}{2a}\right) B\left(\frac{1}{p}, \frac{3}{2}\right) \right.$$

$$+ \frac{\kappa}{B} F\left(\frac{1}{p}, \frac{1}{p}; \frac{1}{2} + \frac{1}{p}; \frac{a-1}{2a}\right) B\left(\frac{1}{p}, \frac{1}{2}\right)$$

$$- \frac{A^{2p}}{a B(p+1)} F\left(1 + \frac{1}{p}, 1 + \frac{1}{p}; \frac{3}{2} + \frac{1}{p}; \frac{a-1}{2a}\right) B\left(\frac{p+1}{p}, \frac{1}{2}\right)$$

$$\left. - \frac{v A^{4p}}{2a^2 B(2p+1)} F\left(2 + \frac{1}{p}, 2 + \frac{1}{p}; \frac{5}{2} + \frac{1}{p}; \frac{a-1}{2a}\right) B\left(\frac{2p+1}{p}, \frac{1}{2}\right) \right] \quad (6.19)$$

In this chapter, the perturbed NLSE with dual-power law nonlinearity that is going to be studied is

$$iq_t + \frac{1}{2}q_{xx} + (|q|^{2p} + v|q|^{4p})q = i\epsilon R[q, q^*] \quad (6.20)$$

where

$$R = \delta|q|^{2m}q + \sigma q \int_{-\infty}^{x} |q|^2 ds \quad (6.21)$$

Now, in the presence of the perturbation term in (6.21), the integrals of motion are modified to give the modified integrals of motion. Using the first two integrals of motion, namely (6.17) and (6.18), the adiabatic variation of the soliton parameters are

$$\frac{dA}{dt} = \frac{\epsilon}{pL A^{p-1}} \left(\frac{p+1}{2p^2}\right)^{\frac{1}{2}} \int_{-\infty}^{\infty} (q^*R + q R^*)dx \quad (6.22)$$

$$\frac{dB}{dt} = \frac{\epsilon}{L} \int_{-\infty}^{\infty} (q^*R + q R^*)dx \quad (6.23)$$

and

$$\frac{d\kappa}{dt} = \frac{\epsilon}{E} \left[i \int_{-\infty}^{\infty} (q_x^*R - q_x R^*)dx - \kappa \int_{-\infty}^{\infty} (q^*R + q R^*)dx \right] \quad (6.24)$$

where E is the energy as given by (6.17) while

$$L = \frac{\Gamma(\frac{1}{2})\Gamma(\frac{1}{p})}{\Gamma(\frac{1}{p} + \frac{1}{2})} \left[\frac{(a-1)(2p+1)}{2v(1+p)} \right]^{\frac{1}{p}}$$

$$\left\{ \frac{2v^2}{ap^3} \frac{(p+1)^3}{(a-1)(2p+1)^2} F\left(\frac{1}{2}, \frac{1}{p}; \frac{1}{2} + \frac{1}{p}; \frac{1-a}{1+a}\right) \right.$$

$$-\frac{2}{B^2}F\left(\frac{1}{2}, \frac{1}{p}; \frac{1}{2}+\frac{1}{p}; \frac{1-a}{1+a}\right)$$

$$-\frac{2v}{ap^2}\frac{(p+1)^2}{(a-1)^2(p+2)(2p+1)}F\left(\frac{1}{2}, \frac{1}{p}; \frac{1}{2}+\frac{1}{p}; \frac{1-a}{1+a}\right)\right\} \quad (6.25)$$

The adiabatic evolution of the energy and linear momentum, in the presence of perturbation terms, by virtue of (2.51) and (2.52) are

$$\frac{dE}{dt} = 2\epsilon \left[\frac{2\delta\,A^{2m+1}}{Ba^{\frac{m+1}{p}}2^{\frac{m+1}{p}}}F\left(\frac{m+1}{p}, \frac{m+1}{p}, \frac{m+1}{p}+\frac{1}{2}; \frac{a-1}{2a}\right)B\left(\frac{m+1}{p}, \frac{1}{2}\right)\right.$$

$$\left.+\frac{\sigma\,A^4}{B^2}\int_{-\infty}^{\infty}\frac{1}{(1+a\cosh\tau)^{\frac{1}{p}}}\left(\int_{-\infty}^{\tau}\frac{1}{(1+a\cosh s)^{\frac{1}{p}}}ds\right)d\tau\right] \quad (6.26)$$

and

$$\frac{dM}{dt} = 2\epsilon\kappa \left[\frac{2\delta\,A^{2m+1}}{Ba^{\frac{m+1}{p}}2^{\frac{m+1}{p}}}F\left(\frac{m+1}{p}, \frac{m+1}{p}, \frac{m+1}{p}+\frac{1}{2}; \frac{a-1}{2a}\right)B\left(\frac{m+1}{p}, \frac{1}{2}\right)\right.$$

$$\left.+\frac{\sigma\,A^4}{B^2}\int_{-\infty}^{\infty}\frac{1}{(1+a\cosh\tau)^{\frac{1}{p}}}\left(\int_{-\infty}^{\tau}\frac{1}{(1+a\cosh s)^{\frac{1}{p}}}ds\right)d\tau\right] \quad (6.27)$$

In order to obtain the adiabatic evolution of the soliton parameters, one can use (6.22)–(6.24) with the perturbation terms given by (6.21) and the form of the dual-power law soliton given by (6.7) to get

$$\frac{dA}{dt} = \frac{2\epsilon}{pL}\frac{A^{3-p}}{B}\left(\frac{p+1}{2p^2}\right)^{\frac{1}{2}}$$

$$\left[\frac{\delta\,A^{2m}}{a^{\frac{m+1}{p}}2^{\frac{m+1}{p}}}F\left(\frac{m+1}{p}, \frac{m+1}{p}, \frac{m+1}{p}+\frac{1}{2}; \frac{a-1}{2a}\right)B\left(\frac{m+1}{p}, \frac{1}{2}\right)\right.$$

$$\left.+\sigma\frac{A^2}{B}\int_{-\infty}^{\infty}\frac{1}{(1+a\cosh\tau)^{\frac{1}{p}}}\left(\int_{-\infty}^{\tau}\frac{1}{(1+a\cosh s)^{\frac{1}{p}}}ds\right)d\tau\right] \quad (6.28)$$

$$\frac{dB}{dt} = \frac{2\epsilon\,A^2}{BL}\left[\frac{\delta\,A^{2m}}{a^{\frac{m+1}{p}}2^{\frac{m+1}{p}}}F\left(\frac{m+1}{p}, \frac{m+1}{p}, \frac{m+1}{p}+\frac{1}{2}; \frac{a-1}{2a}\right)B\left(\frac{m+1}{p}, \frac{1}{2}\right)\right.$$

$$\left.+\sigma\frac{A^2}{B}\int_{-\infty}^{\infty}\frac{1}{(1+a\cosh\tau)^{\frac{1}{p}}}\left(\int_{-\infty}^{\tau}\frac{1}{(1+a\cosh s)^{\frac{1}{p}}}ds\right)d\tau\right] \quad (6.29)$$

and

$$\frac{d\kappa}{dt} = 0 \quad (6.30)$$

The velocity of the soliton, as an evolution of the center of mass, is given by

$$v = \frac{d\bar{x}}{dt} = -\kappa + \frac{\epsilon}{E} \int_{-\infty}^{\infty} x(q^*R + q\,R^*)dx \tag{6.31}$$

which, after integration, or by directly using (2.67), reduces to

$$v = \frac{d\bar{x}}{dt} = -\kappa + 2\epsilon\sigma \frac{A^4}{B^3 E} \int_{-\infty}^{\infty} \frac{\tau}{(1 + a\cosh\tau)^{\frac{1}{p}}} \left(\int_{-\infty}^{\tau} \frac{1}{(1 + a\cosh s)^{\frac{1}{p}}} ds \right) d\tau \tag{6.32}$$

6.4   Quasi-Stationary Solution

For the QS solution to the NLSE with dual-power law nonlinearity, follow the developments of section 2.5. One gets at the leading order

$$-\left\{ \rho_T^{(0)} + \frac{1}{2}(\rho_X^{(0)})^2 \right\} \hat{q}^{(0)} + \frac{1}{2}\frac{\partial^2 \hat{q}^{(0)}}{\partial\theta^2} + (\hat{q}^{(0)})^{2p+1} + v(\hat{q}^{(0)})^{4p+1} = 0 \tag{6.33}$$

and

$$\left(\rho_X^{(0)} - v^{(0)} \right) \frac{\partial \hat{q}^{(0)}}{\partial\theta} = 0 \tag{6.34}$$

Now, (6.34) implies

$$\rho_X^{(0)} = v^{(0)} \tag{6.35}$$

On setting

$$\frac{B^2}{4p^2} = \rho_T^{(0)} + \frac{1}{2}(\rho_X^{(0)})^2 = \rho_T^{(0)} + \frac{1}{2}(v^{(0)})^2 \tag{6.36}$$

(6.33) changes to

$$-\frac{B^2}{4p^2}\hat{q}^{(0)} + \frac{1}{2}\frac{\partial^2 \hat{q}^{(0)}}{\partial\theta^2} + (\hat{q}^{(0)})^{2p+1} + v(\hat{q}^{(0)})^{4p+1} = 0 \tag{6.37}$$

whose solution is

$$\hat{q}^{(0)} = Ag[B(\theta - \bar{\theta})] \tag{6.38}$$

where

$$g(\tau) = \frac{1}{(1 + a\cosh\tau)^{\frac{1}{2p}}} \tag{6.39}$$

and

$$B = A^p \left(\frac{4p^2}{1+p} \right)^{\frac{1}{2p}} \tag{6.40}$$

along with

$$\tau = B(\theta - \bar{\theta}) \tag{6.41}$$

while

$$\frac{d\bar{\theta}}{dt} = v \tag{6.42}$$

At $O(\epsilon)$ level, decomposing $\hat{q}^{(1)} = \hat{\phi}^{(1)} + i\hat{\psi}^{(1)}$ into its real and imaginary parts, the equations for $\hat{\phi}^{(1)}$ and $\hat{\psi}^{(1)}$, by virtue of (6.37), are, respectively,

$$-\frac{B^2}{4p^2} \hat{\phi}^{(1)} + \frac{1}{2} \frac{\partial^2 \hat{\phi}^{(1)}}{\partial \theta^2} + \hat{\phi}^{(1)} \{(2p+1)(\hat{q}^{(0)})^{2p} + v(4p+1)(\hat{q}^{(0)})^{4p}\}$$

$$= \{\rho_T^{(1)} + v^{(0)} \rho_X^{(1)}\} \hat{q}^{(0)} - \frac{\partial^2 \hat{q}^{(0)}}{\partial \theta \partial X} \tag{6.43}$$

and

$$-\frac{B^2}{4p^2} \hat{\psi}^{(1)} + \frac{1}{2} \frac{\partial^2 \hat{\psi}^{(1)}}{\partial \theta^2} + \{(\hat{q}^{(0)})^{2p} + v(\hat{q}^{(0)})^{4p}\} \hat{\psi}^{(1)} = -\frac{\partial \hat{q}^{(0)}}{\partial T} - v^{(0)} \frac{\partial \hat{q}^{(0)}}{\partial X}$$

$$-\{\rho_X^{(1)} - v^{(1)} + \sigma\} \frac{\partial \hat{q}^{(0)}}{\partial \theta} - \rho_{XX}^{(0)} \hat{q}^{(0)} + \delta(\hat{q}^{(0)})^{2m+1}$$

$$+ \sigma \hat{q}^{(0)} \int_{-\infty}^{x} (\hat{q}^{(0)})^2 ds \tag{6.44}$$

In (6.44), setting $\rho_{XX}^{(0)} = 0$ to eliminate frequency chirp gives

$$-\frac{B^2}{4p^2} \hat{\psi}^{(1)} + \frac{1}{2} \frac{\partial^2 \hat{\psi}^{(1)}}{\partial \theta^2} + \{(\hat{q}^{(0)})^{2p} + v(\hat{q}^{(0)})^{4p}\} \hat{\psi}^{(1)} = -\frac{\partial \hat{q}^{(0)}}{\partial T} - v^{(0)} \frac{\partial \hat{q}^{(0)}}{\partial X}$$

$$-\{\rho_X^{(1)} - v^{(1)} + \sigma\} \frac{\partial \hat{q}^{(0)}}{\partial \theta} + \delta(\hat{q}^{(0)})^{2m+1} + \sigma \hat{q}^{(0)} \int_{-\infty}^{x} (\hat{q}^{(0)})^2 ds \tag{6.45}$$

Fredholm's alternative applied to (6.43) yields

$$\frac{\partial B}{\partial X} = 0 \tag{6.46}$$

and

$$\rho_T^{(1)} + v^{(0)} \rho_X^{(1)} = 0 \tag{6.47}$$

whereas, if applied to (6.45), implies

$$\frac{dB}{dT} = \frac{2A^2}{BL} \left[\frac{\delta A^{2m}}{a^{\frac{m+1}{p}} 2^{\frac{m+1}{p}}} F\left(\frac{m+1}{p}, \frac{m+1}{p}, \frac{m+1}{p} + \frac{1}{2}; \frac{a-1}{2a}\right) B\left(\frac{m+1}{p}, \frac{1}{2}\right) \right.$$

$$\left. + \sigma \frac{A^2}{B} \int_{-\infty}^{\infty} \frac{1}{(1 + a \cosh \tau)^{\frac{1}{p}}} \left(\int_{-\infty}^{\tau} \frac{1}{(1 + a \cosh s)^{\frac{1}{p}}} ds \right) d\tau \right] \tag{6.48}$$

and

$$\rho_X^{(1)} = v^{(1)} - \sigma \tag{6.49}$$

Equation (6.46) shows that B is a function of T only and so is A by virtue of (6.40) so that

$$\frac{dA}{dT} = \frac{2}{pL} \frac{A^{3-p}}{B} \left(\frac{p+1}{2p^2}\right)^{\frac{1}{2}} \left[\frac{\delta A^{2m}}{a^{\frac{m+1}{p}} 2^{\frac{m+1}{p}}} F\left(\frac{m+1}{p}, \frac{m+1}{p}, \frac{m+1}{p} + \frac{1}{2}; \frac{a-1}{2a}\right) \right.$$

$$\cdot B\left(\frac{m+1}{p}, \frac{1}{2}\right) + \sigma \frac{A^2}{B} \int_{-\infty}^{\infty} \frac{1}{(1 + a \cosh \tau)^{\frac{1}{p}}}$$

$$\left. \cdot \left(\int_{-\infty}^{\tau} \frac{1}{(1 + a \cosh s)^{\frac{1}{p}}} ds \right) d\tau \right] \tag{6.50}$$

Although (6.48) and (6.50) were obtained before by the soliton perturbation theory (SPT), relations (6.46), (6.47), and (6.49) cannot be recovered by the SPT. Thus, these $O(\epsilon)$ equations reduce to

$$-\frac{B^2}{4p^2} \hat{\phi}^{(1)} + \frac{1}{2} \frac{\partial^2 \hat{\phi}^{(1)}}{\partial \theta^2} + \hat{\phi}^{(1)} \{(2p+1)(\hat{q}^{(0)})^{2p} + v(4p+1)(\hat{q}^{(0)})^{4p}\} = 0 \tag{6.51}$$

and

$$-\frac{B^2}{4p^2} \hat{\psi}^{(1)} + \frac{1}{2} \frac{\partial^2 \hat{\psi}^{(1)}}{\partial \theta^2} + \{(\hat{q}^{(0)})^{2p} + v(\hat{q}^{(0)})^{4p}\} \hat{\psi}^{(1)}$$

$$= -\frac{\partial \hat{q}^{(0)}}{\partial T} + \delta(\hat{q}^{(0)})^{2m+1} + \sigma \hat{q}^{(0)} \int_{-\infty}^{x} (\hat{q}^{(0)})^2 ds \tag{6.52}$$

whose solutions are, respectively,

$$\hat{\phi}^{(1)} = 0 \tag{6.53}$$

and

$$\hat{\psi}^{(1)} - = \frac{A}{Bp} \frac{\partial \bar{\theta}}{\partial T} \frac{1}{(1 + a \cosh \tau)^{\frac{1}{2p}}} \int^{\tau} (1 + a \cosh s_2)^{\frac{1}{p}}$$

$$\cdot \left(\int^{s_2} \frac{a \sinh s_1}{(1 + a \cosh s_1)^{\frac{p+1}{p}}} ds_1 \right) ds_2$$

$$+ \frac{A}{B^3 p} \frac{dB}{dT} \frac{1}{(1 + a \cosh \tau)^{\frac{1}{2p}}} \int^{\tau} (1 + a \cosh s_2)^{\frac{1}{p}}$$

$$\cdot \left(\int^{s_2} \frac{a s_1 \sinh s_1}{(1 + a \cosh s_1)^{\frac{p+1}{p}}} ds_1 \right) ds_2 - \frac{2A}{B^2} \frac{dA}{dT} \frac{1}{(1 + a \cosh \tau)^{\frac{1}{2p}}}$$

$$\cdot \int^{\tau} (1 + a \cosh s_2)^{\frac{1}{p}} \left(\int^{s_2} \frac{1}{(1 + a \cosh s_1)^{\frac{1}{p}}} ds_1 \right) ds_2$$

$$+ 2\delta \frac{A^{2m}}{B^2} \frac{1}{(1 + a \cosh \tau)^{\frac{1}{2p}}} \int^{\tau} (1 + a \cosh s_2)^{\frac{1}{p}}$$

$$\cdot \left(\int^{s_2} \frac{1}{(1 + a \cosh s_1)^{\frac{m+1}{p}}} ds_1 \right) ds_2$$

$$+ 2\sigma \frac{A^2}{B^3} \frac{1}{(1 + a \cosh \tau)^{\frac{1}{2p}}} \int^{\phi} (1 + a \cosh s_3)^{\frac{1}{p}}$$

$$\cdot \left(\int^{s_3} \frac{1}{(1 + a \cosh s_2)^{\frac{1}{p}}} \left(\int_{-\infty}^{s_2} \frac{1}{(1 + a \cosh s_1)^{\frac{1}{p}}} ds_1 \right) ds_2 \right) ds_3 \qquad (6.54)$$

which lead to the QS solution given by (2.98).

The dynamical system in (6.28) and (6.30) or (6.29) and (6.30) has a stable fixed point so that the QS soliton for dual-power law nonlinearity that propagates through an optical fiber with a velocity given by (6.32).

Exercises

1. Prove that for $v \longrightarrow 0$, the soliton given by (6.7) approaches the soliton for power law nonlinearity.

2. Prove that for $v \longrightarrow 0$, the energy of the unperturbed dual-power law soliton reduces to the case of power law solitons—in other words

$$\lim_{v \to 0} E = A^{2-p} \left(\frac{1+p}{2p^2} \right)^{\frac{1}{2}} \frac{\Gamma(\frac{1}{2})\Gamma(\frac{1}{p})}{\Gamma(\frac{1}{p} + \frac{1}{2})}$$

3. Prove that for $p \longrightarrow 1$, the energy of the unperturbed dual-power law soliton reduces to the case of parabolic law solitons—in other

words

$$\lim_{p \to 1} E = \begin{cases} \sqrt{\frac{3}{2v}} \tan^{-1}[2A\sqrt{\frac{v}{3}}] : 0 < v < \infty \\ \sqrt{-\frac{3}{2v}} \tanh^{-1}[2A\sqrt{-\frac{v}{3}}] : -\frac{3}{4A^2} < v < 0 \end{cases}$$

4. Consider the NLSE for dual-power law nonlinearity with filters

$$iq_t + \frac{1}{2}q_{xx} + (|q|^{2p} + v|q|^{4p})q = i\epsilon\beta q_{xx}$$

Prove that the adiabatic variation of the amplitude and frequency are respectively given by

$$\frac{dA}{dt} = -\frac{4\epsilon\beta A^2}{Ba^{\frac{1}{p}}2^{\frac{1}{p}}} \left[\kappa^2 F\left(\frac{1}{p}, \frac{1}{p}, \frac{1}{2} + \frac{1}{p}; \frac{a-1}{2a}\right) B\left(\frac{1}{p}, \frac{1}{2}\right) \right.$$

$$\left. + B^2 F\left(2 + \frac{1}{p}, \frac{1}{p}, \frac{3}{2} + \frac{1}{p}; \frac{a-1}{2a}\right) B\left(\frac{1}{p}, \frac{3}{2}\right) \right]$$

and

$$\frac{d\kappa}{dt} = -\frac{\epsilon\beta\kappa B^2}{4p^2 A^2} \frac{F\left(2 + \frac{1}{p}, 1 + \frac{1}{p}, 2 + \frac{1}{p}; \frac{a-1}{2a}\right)}{F\left(\frac{1}{p}, \frac{1}{p}, \frac{1}{2} + \frac{1}{p}; \frac{a-1}{2a}\right)} \frac{B\left(1 + \frac{1}{p}, 1\right)}{B\left(\frac{1}{p}, \frac{1}{2}\right)}$$

5. For the NLSE with Hamiltonian-type perturbations given by

$$iq_t + \frac{1}{2}q_{xx} + (|q|^{2p} + v|q|^{4p})q = i\epsilon[\lambda(|q|^2 q)_x + \mu(|q|^2)_x q - \gamma q_{xxx}]$$

prove that the velocity is given by

$$v = -\kappa - \epsilon(\mu + 3\gamma B^2) - 3\frac{\epsilon\gamma B^2}{2p^2} \frac{F\left(2 + \frac{1}{p}, \frac{1}{p}, \frac{1}{p} + \frac{3}{2}; \frac{a-1}{2a}\right)}{F\left(\frac{1}{p}, \frac{1}{p}, \frac{1}{2} + \frac{1}{p}; \frac{a-1}{2a}\right)} \frac{B\left(\frac{1}{p}, \frac{3}{2}\right)}{B\left(\frac{1}{p}, \frac{1}{2}\right)}$$

$$- \epsilon(3\lambda + 2\mu) \frac{A^2}{2^{\frac{1}{p}}a^{\frac{1}{p}}} \frac{F\left(\frac{2}{p}, \frac{2}{p}, \frac{2}{p} + \frac{1}{2}; \frac{a-1}{2a}\right)}{F\left(\frac{1}{p}, \frac{1}{p}, \frac{1}{2} + \frac{1}{p}; \frac{a-1}{2a}\right)} \frac{B\left(\frac{2}{p}, \frac{1}{2}\right)}{B\left(\frac{1}{p}, \frac{1}{2}\right)}$$

7

Saturable Law Nonlinearity

An important class of optical nonlinearity that has recently drawn much attention is the saturable law nonlinearity, also known as *saturating nonlinearity*. This chapter is devoted to the study of soliton propagation in saturating media. Section 7.1 explains the basic features of saturating nonlinearity. The derivation of the nonlinear Schrödinger's equation (NLSE) with saturable law is given in Section 7.2. The conserved quantities are identified in Section 7.3. Saturating media admit bistable solitons, which are discussed in Section 7.4. Finally, Section 7.5 is devoted to the discussion of arbitrary pulse propagation.

7.1 Introduction

The practical interest in the investigation of optical pulse propagation in nonlinear waveguides is concentrated on low-loss materials with nonresonant nonlinearities. At low intensity of optical field nonresonant nonlinearity in materials of practical interest resembles Kerr nonlinearity. Therefore, a very large number of investigations on temporal soliton propagation in fibers and waveguides have been carried out employing Kerr law nonlinearity. In Kerr media, the magnitude of the induced nonlinearity increases linearly with an increase in the value of the intensity of the optical field. However, as the incident field becomes stronger, for optical fields whose frequencies approach a resonant frequency of the material, non-Kerr higher order nonlinearity comes into play, essentially changing the physical features and stability of optical soliton propagation. Particularly for short pulses and high-input peak power, the field-induced change of the refractive index cannot be described by a Kerr-type nonlinearity because it is influenced by higher order nonlinearities. As a consequence, the optically induced refractive index change becomes saturated at higher field strength. This is especially important in materials with higher nonlinear coefficients, for example, semiconductor-doped glasses and organic polymers in which the saturation of nonlinear refractive index changes come to play at moderately high intensities and should be

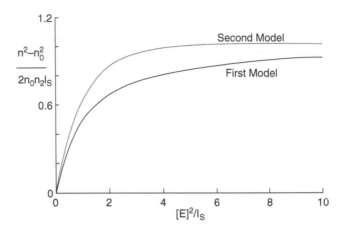

FIGURE 7.1
Behavior of two different form of saturating nonlinearity.

taken into account. The saturating nonlinearity can be modeled by different equations. Two known models expressing the nonlinear refractive index n are

$$n^2 = n_0^2 + \frac{2n_0 n_2 |E|^2}{1 + |E|^2 / I_s} \tag{7.1}$$

and

$$n^2 = n_0^2 + 2n_0 n_2 I_s \left(1 - e^{-\frac{|E|^2}{I_s}} \right) \tag{7.2}$$

where n_0 is, as usual, the linear refractive index; E is the electric field; and n_2 and I_s are respectively the third-order nonlinear coefficients and the characteristic saturation intensity. The parameter n_2, as seen in chapter 3, is also known as the Kerr coefficient. For small intensities, namely when $|E|^2 \ll I_s$, both (7.1) and (7.2) reduce to Kerr law nonlinearity. However, for large intensities, namely when $|E|^2 \gg I_s$, the refractive index saturates and approaches its maximum value, $2n_0 n_2 I_s$. The variation of the induced refractive index with $|E|^2$ is shown in Figure 7.1. Both models signify saturation at large intensity. Since both models produce identical qualitative features of solitons, discussion using either one of them will be sufficient. Hence, one may choose only the first one for further discussion.

7.2   The NLSE

Consider the propagation of an optical pulse in a medium possessing the higher order nonlinear refractive index given by (7.1). The wave equation that describes the propagation of slowly varying envelope $A(\tau, t)$ of the pulse along *t*-direction may be written as

$$i\left(\frac{\partial A}{\partial t} + k_\omega \frac{\partial A}{\partial \tau}\right) - \frac{k_{\omega\omega}}{2}\frac{\partial^2 A}{\partial \tau^2} + \left(\frac{n_2 k}{n_0}\right)\frac{|A|^2}{1 + |A|^2/I_s} + \frac{i\alpha A}{2} = 0 \qquad (7.3)$$

where k is the propagation constant, $k_\omega = \partial k/\partial \omega$ represents the reciprocal of group velocity, and $k_{\omega\omega} = \partial^2 k/\partial \omega^2$ is the group velocity dispersion at the carrier frequency ω of the pulse. The parameter α is the attenuation coefficient. To normalize (7.3), the following transformations are introduced

$$x = \frac{\tau - tk_\omega}{x_0} \qquad (7.4)$$

$$A = u\sqrt{P_0} \qquad (7.5)$$

$$P_0 = \frac{|k_{\omega\omega}(0)|n_0}{n_2 k x_0} \qquad (7.6)$$

$$L_D = \frac{x_0}{|k_{\omega\omega}(0)|} \qquad (7.7)$$

$$s = \frac{P_0}{I_s} \qquad (7.8)$$

where x_0 is the characteristic time scale. In terms of these normalized variables, (7.3) reduces to

$$i\frac{\partial u}{\partial t} + \frac{\sigma}{2}\frac{\partial^2 u}{\partial x^2} + \frac{|u|^2 u}{1 + s|u|^2} + i\Gamma u = 0 \qquad (7.9)$$

where $\Gamma = \alpha n_0/2n_2 k\sqrt{P_0}$ and $\sigma = \pm 1$ depending on the sign of the parameter $k_{\omega\omega}$. A positive sign corresponds to the anomalous dispersion regime $k_{\omega\omega} < 0$, while a negative sign corresponds to the normal dispersion regime $k_{\omega\omega} > 0$. Finally, introducing the transformation $u = Ue^{-\Gamma t}$, one obtains

$$i\frac{\partial U}{\partial t} + \frac{\sigma}{2}\frac{\partial^2 U}{\partial x^2} + \frac{f(t)|U|^2 U}{1 + sf(t)|U|^2} = 0 \qquad (7.10)$$

where $f(t) = e^{-2\Gamma t}$. Equation (7.10) is the modified NLSE (MNLSE) due to the saturable law of nonlinearity. This equation is not integrable by the IST. This implies that any input beam propagating in this medium cannot be decomposed into stable stationary waves and radiation. This fact has a profound effect on the properties of solitons. Particularly, the basic notion that solitons are pulses whose properties remain invariant with propagation is no longer valid. In addition, the contention that solitons undergo elastic collision—that is, if two solitons collide, there is no loss of energy to radiation fields—is no longer valid. In saturable media, soliton shape may change periodically or evolve monotonically and the collision may become inelastic. However, the essential properties of localized propagation are still applicable in saturable media and commonly provide a richer and more useful range of propagation characteristics than Kerr law nonlinearity.

7.2.1   Conserved Quantities

The MNLSE with saturable law nonlinearity has three integrals of motion, similar to the other laws of nonlinearity seen in previous chapters. They are respectively the energy (E), momentum (M), and Hamiltonian (H). For $f(t) = 1$, the three conserved quantities are as follows

$$E = \int_{-\infty}^{\infty} |U|^2 dx \tag{7.11}$$

$$M = \frac{i}{2} \int_{-\infty}^{\infty} (U^* U_x - U U_x^*) dx \tag{7.12}$$

and

$$H = \int_{-\infty}^{\infty} \left[\frac{\sigma}{2} |U_x|^2 - f(I) \right] dx \tag{7.13}$$

where

$$f(I) = \int_0^I \frac{z}{1 + sz} dz \tag{7.14}$$

and the intensity I is given by $I = |U|^2$.

7.3   Bistable Solitons

Here and henceforth, only the anomalous dispersive media (namely $\sigma = 1$) will be considered. Moreover, in this section, the lossless media will be studied by setting $f(t) = 1$. Since (7.10) is not integrable by the IST and no solutions by the traveling wave ansatz are known, one may search its soliton solution by numerical technique. Thus, one may seek solutions of the form [261]

$$U(x, t) = \sqrt{\Psi(x)} e^{i\beta t} \tag{7.15}$$

where β physically represents the nonlinear propagation constant shift. For a bright soliton, one can impose the boundary conditions

$$\lim_{x \to \pm\infty} \Psi(x) = 0 \tag{7.16}$$

and

$$\lim_{x \to \pm\infty} \frac{d\Psi}{dx} = 0 \tag{7.17}$$

Using ansatz (7.15) in (7.10), one obtains the following ordinary differential equation (ODE)

$$\frac{1}{4\Psi}\frac{d^2\Psi}{dx^2} - \frac{1}{8\Psi^2}\left(\frac{d\Psi}{dx}\right)^2 + \frac{\Psi}{1+s\Psi} - \beta = 0 \tag{7.18}$$

This ODE can be solved numerically to obtain the stationary pulse profile for a given nonlinear propagation constant β. However, arbitrary values of β do not permit stationary pulses. Depending on the peak power of the soliton, only certain values of β are permissible. Now (7.18) can be rearranged as

$$\frac{1}{8}\frac{d}{d\Psi}\left(\frac{d\Psi}{dx}\frac{1}{\sqrt{\Psi}}\right)^2 + \frac{\Psi}{1+s\Psi} - \beta = 0 \tag{7.19}$$

which, on integration, yields

$$\frac{1}{8}\left(\frac{d\Psi}{dx}\frac{1}{\sqrt{\Psi}}\right)^2 + \frac{1}{s^2}[s\Psi - \ln(1+s\Psi)] - \beta\Psi = 0 \tag{7.20}$$

Since for a bright soliton $d\Psi/dx = 0$ at the peak of the soliton amplitude (namely at $\Psi = \Psi_0$), one can immediately obtain

$$\beta = \frac{1}{s^2}\left[s - \frac{\ln(1+s\Psi_0)}{\Psi_0}\right] \tag{7.21}$$

Using the above relationship, one can easily determine the permissible values of β for the corresponding given soliton's peak amplitude Ψ_0. Figure 7.2(a) depicts Ψ_0 as a function of β.

Soliton energy can be estimated using

$$E = \int_{-\infty}^{\infty}|U|^2 dx = |\Psi_0|^2 \int_{-\infty}^{\infty}|g(x)|^2 dx \tag{7.22}$$

where $g(x)$ is the soliton shape function. Equation (7.18) can be solved numerically to find out the shape of the soliton, its energy, and its temporal width τ_0. A typical behavior is shown in Figures 7.2(b) and 7.2(c). An interesting feature to note is that E is a multivalued function of τ_0, thus admitting bistable [261] solitons, namely the existence of two solitons having same τ_0 but different energy.

7.4 Arbitrary Pulse Propagation

When a pulse shape is not in the form of a stationary pulse, such as those admissible by (7.18), it can still propagate in a media without radiating out. However, depending on the peak power and width of the pulse, when the pulse propagates, it might breathe or its width might increase monotonically.

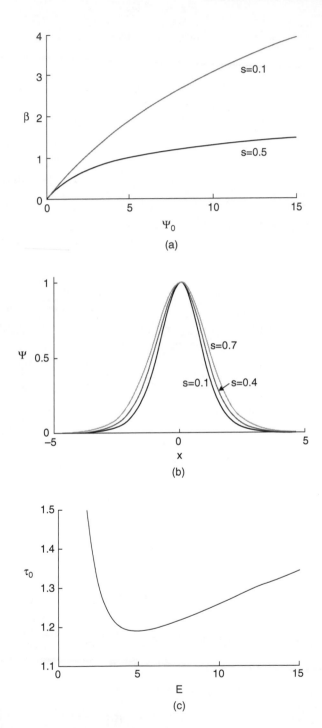

FIGURE 7.2
(a) Variation of β with ψ_0, (b) shape of soliton with different s, and (c) variation of the temporal width with soliton energy E.

The behavior of the pulse depends on the relative magnitude of nonlinearity induced self-phase modulation and fiber dispersion. When the pulse propagates in a medium, it acquires a chirp. The anomalous dispersion produces a chirp of opposite sign. The propagation characteristics of the pulse result as an interplay between these two opposite chirps.

In order to investigate the dynamic behavior, the variational principle that was introduced in chapter 3 will be exploited here. The Lagrangian can be written as

$$L = i\left(U\frac{\partial U^*}{\partial t} - U^*\frac{\partial U}{\partial t}\right) + \left|\frac{\partial U}{\partial x}\right|^2 - \frac{2}{s}|U|^2 + \frac{2}{s^2 f(t)} \ln\left\{1 + sf(t)|U|^2\right\}$$

(7.23)

By Hamilton's principle of least action, one can write

$$\delta \int L dt = 0 \qquad (7.24)$$

In general, the temporal profile of an optical pulse emitted by a present-day semiconductor laser is very close to sech or Gaussian. Thus, it is reasonable that one would be curious to know the behavior of such pulses in saturating nonlinear media. Therefore, for further investigation, one can choose the trial function

$$U(x, t) = \frac{B(t)}{\cosh\left(\frac{x}{a(t)}\right)} e^{i\phi(t)x^2 + \psi(t)}$$

(7.25)

where $B(t), a(t), \phi(t)$, and $\psi(t)$ are the amplitude, duration, chirp, and longitudinal phase of the pulse, respectively. By using the trial function (7.25), the averaged Lagrangian is given by

$$L = 4B^2 a \frac{d\psi}{dt} + \frac{2B^2}{3a} + \frac{\pi^2}{6}\left(4\phi^2 + 2\frac{d\phi}{dt}\right)B^2 a^3 - \frac{4B^2 a}{s}$$

$$+ \frac{2a}{f(t)s^2} \int_{-\infty}^{\infty} \ln\left\{1 + \frac{sf(t)B^2}{\cosh^2 y}\right\} dy \qquad (7.26)$$

Employing the Raleigh-Ritz optimization procedure, one arrives at the following set of equations

$$B^2(t)a(t) = B^2(0)a(0) = constant = N^2 \qquad (7.27)$$

$$\phi(t) = \frac{1}{2a(t)}\frac{da}{dt} \qquad (7.28)$$

$$\frac{d^2 a}{dt^2} = \frac{4}{\pi^2 a^3} - \frac{6}{\pi^2 N^2 f(t)s^2}\frac{d}{da}\{aG(a)\} \qquad (7.29)$$

and

$$\frac{d\psi}{dt} = -\frac{1}{3a^2} + \frac{a}{2N^2s^2 f(t)}\frac{d}{da}\{aG(a)\} + \frac{1}{s} + \frac{a}{4N^2s^2 f(t)}\left(a\frac{\partial G}{\partial a} - G\right)$$

(7.30)

where

$$G(a) = 2\left[\sinh^{-1}\left\{\frac{\sqrt{sf(t)}}{\sqrt{a}}N\right\}\right]^2$$

(7.31)

Equations (7.27)–(7.30) can be used to investigate the propagation of a sech optical pulse in lossless and lossy medium with saturating nonlinearity.

7.4.1 Lossless Uniform Media ($\Gamma = 0$)

In this section, the propagation characteristics of an optical pulse in a lossless medium are investigated. For such a medium, $f(t) = 1$. So (7.29) reduces to

$$\frac{d^2a}{dt^2} = \frac{4}{\pi^2 a^3} - \frac{6}{\pi^2 N^2 s^2}\frac{d}{da}\{aG(a)\}$$

(7.32)

which can be integrated at once to give

$$\frac{1}{2}\left(\frac{da}{dt}\right)^2 + V(a) = 0$$

(7.33)

where

$$V(a) = \frac{2}{\pi^2}\left(\frac{1}{a^2} - \frac{1}{a_0^2}\right) + \frac{6}{\pi^2 N^2 s^2}[aG(a) - a_0G(a_0)] - \frac{a_0^2 C_0^2}{2}$$

(7.34)

with $a_0 = a(t = 0)$ and $C_0 = (da/dt)_{t=0}$, which can be identified as initial chirp at $t = 0$. The left side of (7.33) can be identified as total energy and is identical to that of an oscillator of unit mass that is executing its motion under a potential $V(a)$ and, hence, (7.32) is derivable from the gradient of $V(a)$, namely

$$\frac{d^2a}{dt^2} = -\frac{\partial V(a)}{\partial a}$$

(7.35)

Although the analytical solution of (7.35) is not obvious, a lot of valuable information about a propagating pulse can be extracted by examining the potential $V(a)$.

For small saturation parameter $s \to 0$, and for small N, $N\sqrt{s}/\sqrt{a} \ll 1$; therefore

$$\left[\sinh^{-1}\left\{\frac{\sqrt{s}}{\sqrt{a}}N\right\}\right]^2 \approx \frac{N^2s}{a} + \frac{8}{45}\frac{N^6s^3}{a^3} - \frac{N^4s^2}{3a^2}$$

(7.36)

and the potential reduces to

$$V(a) \approx \frac{2}{\pi^2 a^2} - \frac{4N^7}{\pi^2 a} + \frac{32}{15} \frac{N^4 s}{\pi^2 a^2} + \frac{12}{\pi^2 s} \tag{7.37}$$

Using (7.32), one recovers the governing equation of a sech soliton propagating through a parabolic nonlinear medium

$$\frac{d^2 a}{dt^2} = \frac{4}{\pi^2 a^3} - \frac{4N^2}{\pi^2 a} + \frac{64}{15} \frac{N^4 s}{\pi^2 a^3} \tag{7.38}$$

and, on setting $s = 0$ in (7.38), one recovers the result of a sech soliton propagating through a Kerr nonlinear medium.

The potential $V(a)$ given by (7.34) can be examined for saturating nonlinearity. It is clear from (7.34) that $V(a) \leq 0$ is the allowed propagation region. Thus, for a given initial chirp C_0, the value of N for which bound motion is possible can be obtained from the potential $V(a)$. For very large a, if $V(a)$ remains negative, unbounded motion occurs. Physically, this corresponds to a situation in which the temporal width of the soliton is very large—in other words, the pulse loses its solitonic character. Thus, unbounded motion results in the region where the following condition holds

$$F(N) = 6s - \frac{s^2}{a_0^2} - \frac{6a_0}{N^2} \left[\sinh^{-1} \left(\frac{N\sqrt{s}}{\sqrt{a_0}} \right) \right]^2 - \frac{\pi^2 C_0^2 s^2 a_0^2}{4} < 0 \tag{7.39}$$

For further discussion, one can assume $a_0 = 1$, without any loss of generality. For initial chirpless pulse, $C_0 = 0$, the variation of $F(N)$ with N is shown in Figure 7.3(a) for different values of the saturation parameter s. $F(N) < 0$ is the region of unbounded motion and $F(N) > 0$ is the region of bounded motion. The threshold value of N (N_{th}) above which the bound motion is expected can be estimated by noting down the value of N for which $F(N) = 0$. From Figure 7.3(a) it is evident that N_{th} increases with increases in the value of the saturation parameter s. In Figure 7.3(b), $F(N)$ is plotted against N for different chirp. It is evident from this figure that N_{th} increases with increases in the value of the initial chirp C_0. Thus, one can conclude that indefinite increases in the value of the chirp may lead to destruction of the soliton of given energy N^2. For example, for $C_0 = 0$, $N = 1$ gives a stable soliton propagation. However, when $C_0 = 0.9$, $F(N) < 0$ for $N < 1.2$. Thus the soliton is unstable and would disintegrate.

For a chirpless pulse, variation of $V(a)$ against a for different N and $s = 0.3$ is shown in Figure (7.4). One can identify four distinctly different types of behaviors of the pulse width for different values of the parameter N. These are

1. Stationary propagation with $a = 1$ for $N = 1.226$
2. Oscillatory bounded motion with $0 < a < 1$ for $N > 1.226$
3. Oscillatory bounded motion with $0.73 < N < 1.226$
4. Unbounded motion with $a \to \infty$ for $N < 0.73$.

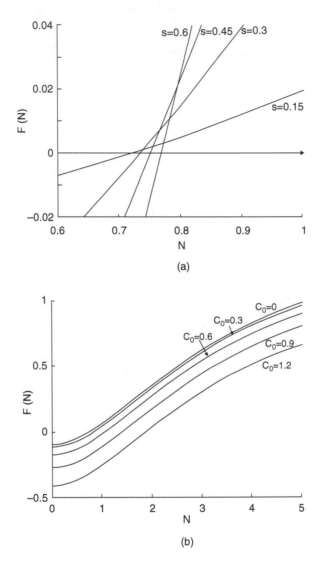

FIGURE 7.3
(a) Variation of F with N for different s, (b) variation of F with N for different initial chirp.

7.4.2 Stationary Pulse Propagation

For a chirpless, stationary pulse, the condition can be obtained by using $\partial V/\partial a = 0$. Thus, the value of N for which a stationary pulse is obtained can be found from the relation

$$N^2 s^2 - 3a^3 \left[a \left\{ \sinh^{-1}\left(\frac{N\sqrt{s}}{\sqrt{a}} \right) \right\}^2 - \left(\frac{N\sqrt{s}}{\sqrt{a}} \right) \frac{\sinh^{-1}\left(\frac{N\sqrt{s}}{\sqrt{a}} \right)}{\left(1 + \frac{N^2 s}{a} \right)^{\frac{1}{2}}} \right] = 0$$

$$(7.40)$$

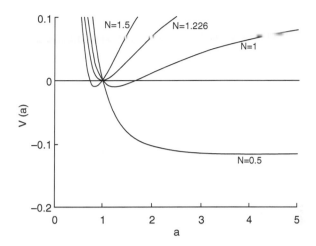

FIGURE 7.4
Variation of potential V with a for different N.

This equation can be solved numerically to find out the value of N for a given a for stationary propagation. Figure (7.5) depicts the variation of a with N for stationary pulse propagation. The existence of two-state soliton is clearly evident from the figure. The stability of these solitons can be found from the sign of $\partial^2 V/\partial a^2$. It was verified that $\partial^2 V/\partial a^2 > 0$, so that these stationary states are stable.

7.4.3 Lossy Media ($\Gamma \neq 0$)

For a lossy medium, $\Gamma \neq 0$ so that $f(t) \neq 1$. Since the right side of (7.29) depends on t explicitly, the potential formalism that was developed in the previous section cannot be used here. Particularly, d^2a/dt^2 is not derivable

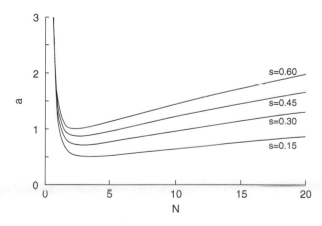

FIGURE 7.5
Variation of soliton width with N for different s.

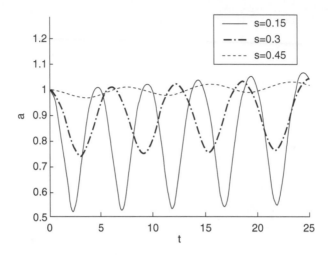

FIGURE 7.6
Variation of beam width with distance of propagation.

from the gradient of a scalar potential. Therefore, one can solve (7.29) numerically to study the influence of dissipation. Equation (7.29) is solved numerically using a chirpless input pulse of width $a_0 = 1$ and normalized dissipation $\Gamma = 0.001$. For a typical value of $N = 1.5$, variation of pulse width is displayed in Figure 7.6. It is evident from the figure that, as the pulse propagates, its temporal width oscillates. As a consequence of energy loss due to dissipation, net broadening of width takes place over many cycles of oscillations. The net broadening decreases with the increase in the value of s. One can conclude from Figure 7.6 that the pulse is more robust under dissipation loss with increasing value of saturation.

Exercises

1. Consider the Lagrangian that is given by (7.23). Now, use the relationship

$$\frac{\delta L}{\delta U^*} = \frac{\delta L}{\delta U} = 0$$

   to recover the NLSE that is given by (7.10).

2. Show that in (7.29) $G(a)$ is given by

$$G(a) = \int_{-\infty}^{\infty} \ln\left[1 + \frac{s N^2 f(t)}{a \cosh^2 y}\right] dy$$

which reduces to

$$G(a) - 2\left[\sinh^{-1}\left\{\frac{\sqrt{\partial f(t)}}{\sqrt{a}} N\right\}\right]^2$$

that is given by (7.31).

3. Set up a coupled NLSE for two incoherently coupled solitons in saturating nonlinear media. Show that the Lagrangian for such a system is given by

$$L = \sum_{j=1}^{2}\left[i\left(\psi_j\frac{\partial\psi_j^*}{\partial t} - \psi^*\frac{\partial\psi}{\partial t}\right) + \left|\frac{\partial\psi_j}{\partial x}\right|^2 - 2\left|\psi_j\right|^2\right]$$

$$+ 2\ln\left(1 + \sum_{j=1}^{2}\left|\psi_j\right|^2\right)$$

4. Use Euler's equation to derive (7.29) and (7.30).

5. Recall that the NLSE for a parabolic nonlinear media is given by

$$i\frac{\partial\psi}{\partial t} + \frac{1}{2}\frac{\partial^2\psi}{\partial x^2} + \left(|\psi|^2 + v\,|\psi|^4\right)\psi = 0$$

Consider the input pulse that is given by

$$\psi(x, t) = \frac{A(t)}{\cosh\left(\frac{x}{B}\right)}$$

Now, prove that the width $B(t)$ will vary according to

$$\frac{d^2 B}{dt^2} = \frac{4}{\pi^2 B^2} - \frac{4N^2}{\pi^2 B} + \frac{64v N^4}{15\pi^2 B^3}$$

where $N^2 = A^2 B = $ constant.

8

Soliton–Soliton Interaction

The intrachannel collision of optical solitons in the presence of perturbation terms is studied in this chapter by the aid of quasi-particle theory (QPT) that will be developed from basic principles. The nonlinearities that are studied in this chapter are the Kerr, power, parabolic, and dual-power laws. The perturbation terms that are considered in this chapter are both Hamiltonian and non-Hamiltonian type.

8.1 Introduction

In a soliton communication system, it is necessary to launch the solitons close to each other to enhance the information-carrying capacity of the fiber. Two solitons that are placed too close to each other can mutually interact, thus providing a very serious hindrance to the performance of the soliton transmission system. However, the presence of the perturbation terms of the nonlinear Schrödinger's equation (NLSE) can lead to the suppression of the two-soliton interaction, thus solving this problem [62, 185].

A considerable amount of research effort has been spent on the reduction of interaction of solitons. For example, the use of Gaussian-shaped pulses reduces the interaction because of the steep slope, but this is achieved at the expense of creating larger oscillatory tails. It has also been shown that introducing a phase difference between neighboring solitons can lead to a reduction in their interaction. Incoherent interaction of solitons has also been analyzed. The third-order dispersion of an optical fiber can also be used to reduce mutual interactions, but this results in the breakup of the bound state of solitons. A more realistic way to reduce these interactions is to launch adjacent pulses with unequal amplitudes. In this case, solitons form a bound system and effectively maintain their initial pulse separation. The higher order nonlinear effect is also shown to break up the bound state of the solitons. A detailed study of soliton–soliton interaction in the presence of fiber loss and periodic amplification has also been carried out [185].

In this chapter, the QPT will be introduced and its use as a mechanism for suppressing the interaction of solitons due to the presence of perturbation terms will be studied. The QPT will be formulated for Kerr law, power law, parabolic law, and dual-power law. Numerical simulations will also support the theory.

8.2 Mathematical Formulation

To study the intrachannel collision of optical solitons, the NLSE with the following perturbation terms is considered

$$iq_t + \frac{1}{2}q_{xx} + F(|q|^2)q = i\epsilon R[q, q^*] \tag{8.1}$$

where

$$R = \delta|q|^{2m}q - \alpha q_x + \beta q_{xx} - \gamma q_{xxx}$$
$$+ \sigma q \int_{-\infty}^{x} |q|^2 d\xi + \lambda(|q|^2 q)_x + \mu(|q|^2)_x q - i\rho q_{xxxx} \tag{8.2}$$

The coefficient of β is called the *bandpass filtering term*. Also, in (8.2), α is the frequency separation between the soliton carrier and the frequency at the peak of EDFA gain. Moreover, λ is the self-steepening coefficient for short pulses (typically ≤ 100 femto seconds), μ is the nonlinear dispersion coefficient, and γ is the coefficient of the third-order dispersion [241], while ρ represents the coefficient of fourth-order dispersion [64]. It needs to be noted that the coefficients δ, β, and σ represent non-Hamiltonian perturbations, while the remaining terms in (8.2) represent Hamiltonian-type perturbations [86].

The soliton solution of (8.1) for $\epsilon = 0$, although not integrable, is assumed to be given in the form

$$q(x, t) = \eta(t)g[\zeta(t)(x - vt - x_0)]e^{(-i\kappa x + i\omega t + i\sigma_0)} \tag{8.3}$$

where

$$\kappa = -v \tag{8.4}$$

$$\zeta(t) = \chi(\eta(t)) \tag{8.5}$$

and

$$\omega(t) = \psi(\eta(t), \kappa(t)) \tag{8.6}$$

In (8.5), g represents the shape of the soliton described by the NLSE and depends on the type of nonlinearity in (8.1). The parameters $\eta(t)$ and $\zeta(t)$ in (8.5) respectively represent the soliton amplitude and width, while $\kappa(t)$ and $\omega(t)$ are the frequency and wave number of the soliton, respectively, and v

is the velocity. The explicit functional form of ψ in (8.6) is actually given in (2.39). Also, x_0 and σ_0 respectively represent the center of the soliton and the center of the soliton phase. In (8.5) and (8.6), the functional forms χ and ψ depend on the type of nonlinearity in (8.1).

The 2-soliton solution of the NLSE (8.1) takes the asymptotic form

$$q(x, t) = \sum_{l=1}^{2} \eta_l(t) g\left[\zeta_l(t)(x - v_l t - x_l)\right] e^{(-i\kappa_l x + i\omega_l t + i\sigma_l)} \tag{8.7}$$

with

$$\kappa_l = -v_l \tag{8.8}$$

$$\zeta_l(t) = \chi(\eta_l(t)) \tag{8.9}$$

$$\omega_l(t) = \psi(\eta_l(t), \kappa_l(t)) \tag{8.10}$$

where $l = 1, 2$. In the study of soliton–soliton interaction (SSI), the initial pulse waveform is taken to be of the form

$$q(x, 0) = \eta_1 g\left[\zeta_1\left(x - \frac{x_0}{2}\right)\right] e^{i\phi_1} + \eta_2 g\left[\zeta_2\left(x + \frac{x_0}{2}\right)\right] e^{i\phi_2} \tag{8.11}$$

which represents the injection of 2-soliton-like pulses into a fiber. Here, x_0 represents the initial separation of the solitons, namely the center-to-center soliton separation. It is to be noted that for $x_0 \to \infty$, (8.11) represents exact soliton solutions, but for $x_0 \sim O(1)$, it does not represent an exact 2-soliton solution. The initial pulse form is modified depending on the type of perturbation considered, as seen below.

1. *Non-Hamiltonian Perturbations:* For studying the SSI with non-Hamiltonian perturbations, the case of in-phase injection of solitons with equal amplitudes will be considered. So, without any loss of generality, $\eta_1 = \eta_2 = 1$ and $\phi_1 = \phi_2 = 0$ are chosen so that (8.11) modifies to

$$q(x, 0) = g\left[\zeta_1\left(x - \frac{x_0}{2}\right)\right] + g\left[\zeta_2\left(x + \frac{x_0}{2}\right)\right] \tag{8.12}$$

2. *Hamiltonian Perturbations:* For studying SSI with Hamiltonian perturbations, the case of in-phase injection of solitons with unequal amplitudes will be considered. So, without any loss of generality, $\eta_1 = \eta_0, \eta_2 = 1$ and $\phi_1 = \phi_2 = 0$ are chosen so that (8.11) modifies to

$$q(0, T) = \eta_0 g\left[\zeta_0\left(x - \frac{x_0}{2}\right)\right] + g\left[\zeta\left(x + \frac{x_0}{2}\right)\right] \tag{8.13}$$

where

$$\zeta_0 = \chi(\eta_0) \tag{8.14}$$

and ζ is given by (8.5).

The special cases with regards to the four laws of nonlinearity will now be individually discussed in the following four subsections.

8.2.1   Kerr Law

For the case of Kerr law nonlinearity, $F(s) = s$ so that (8.1) becomes

$$iq_t + \frac{1}{2}q_{xx} + |q|^2 q = i\epsilon R[q, q^*] \tag{8.15}$$

The 1-soliton solution of (8.15) for $\epsilon = 0$ that can be obtained by the inverse scattering transform (IST) has the form

$$q(x, t) = \frac{\eta}{\cosh[\zeta(x - vt - x_0)]} e^{i(-\kappa x + \omega t + \sigma_0)} \tag{8.16}$$

where

$$\zeta \equiv \chi(\eta) = \eta \tag{8.17}$$

and

$$\omega \equiv \psi(\eta, \kappa) = \frac{\eta^2 - \kappa^2}{2} \tag{8.18}$$

Also, the 2-soliton solution of the NLSE (8.1) takes the asymptotic form

$$q(x, t) = \sum_{l=1}^{2} \frac{\eta_l}{\cosh\left[\eta_l(x - v_l t - x_{0_l})\right]} e^{i\left(-\kappa_l x + \omega_l t + \sigma_{0_l}\right)} \tag{8.19}$$

where

$$\zeta_l \equiv \chi(\eta_l) = \eta_l \tag{8.20}$$

and

$$\omega_l \equiv \psi(\eta_l, \kappa_l) = \frac{\eta_l^2 - \kappa_l^2}{2} \tag{8.21}$$

and $l = 1, 2$. In the study of SSI with non-Hamiltonian perturbations, the initial pulse form is taken to be

$$q(x, 0) = \frac{\eta_1}{\cosh\left[\eta_1\left(x - \frac{x_0}{2}\right)\right]} e^{i\phi_1} + \frac{\eta_2}{\cosh\left[\eta_2\left(x + \frac{x_0}{2}\right)\right]} e^{i\phi_2} \tag{8.22}$$

For non-Hamiltonian perturbations, the choice $\eta_1 = \eta_2 = 1$ and $\phi_1 = \phi_2 = 0$ gives

$$q(x, 0) = \frac{1}{\cosh\left[\left(x - \frac{x_0}{2}\right)\right]} + \frac{1}{\cosh\left(x + \frac{x_0}{2}\right)} \tag{8.23}$$

which represents an in-phase injection of pulses with equal amplitudes.

For Hamiltonian-type perturbations, the choice $\eta_1 = \eta_0$, $\eta_2 = 1$, and $\phi_1 = \phi_2 = 0$ gives

$$q(x, 0) = \frac{\eta_0}{\cosh\left[\eta_0\left(x - \frac{x_0}{2}\right)\right]} + \frac{1}{\cosh\left(x + \frac{x_0}{2}\right)} \tag{8.24}$$

which represents an in-phase injection of pulses with unequal amplitudes.

8.2.2 Power Law

For the case of power law nonlinearity, the NLSE is given by

$$iq_t + \frac{1}{2}q_{xx} + |q|^{2p}q = i\epsilon R[q, q^*] \tag{8.25}$$

In this case, (8.25) with $\epsilon = 0$ is not integrable, as seen before, by the IST. However, (8.25) for $\epsilon = 0$ supports solitons of the form

$$q(x, t) = \frac{\eta}{\cosh^{\frac{1}{p}}\left[\zeta(x - vt - x_0)\right]}e^{i(-\kappa x + \omega t + \sigma_0)} \tag{8.26}$$

where

$$\zeta \equiv \chi(\eta) = \eta^p\left(\frac{2p^2}{1+p}\right)^{\frac{1}{2}} \tag{8.27}$$

and

$$\omega \equiv \psi(\eta, \kappa) = \frac{\zeta^2}{2p^2} - \frac{\kappa^2}{2} \tag{8.28}$$

The 2-soliton solution of the NLSE takes the asymptotic form

$$q(x, t) = \sum_{l=1}^{2}\frac{\eta_l}{\cosh^{\frac{1}{p}}\left[\zeta_l(x - v_l t - x_{0_l})\right]}e^{i\left(-\kappa_l x + \omega_l t + \sigma_{0_l}\right)} \tag{8.29}$$

where

$$\zeta_l \equiv \chi(\eta_l) = \eta_l^p\left(\frac{2p^2}{1+p}\right)^{\frac{1}{2}} \tag{8.30}$$

and

$$\omega_l \equiv \psi(\eta_l, \kappa_l) = \frac{\zeta_l^2}{2p^2} - \frac{\kappa_l^2}{2} \tag{8.31}$$

In the study of SSI for power law, the initial pulse waveform is assumed to be

$$q(x, 0) = \frac{\eta_1}{\cosh^{\frac{1}{p}}\left[\zeta_1\left(x - \frac{x_0}{2}\right)\right]}e^{i\phi_1} + \frac{\eta_2}{\cosh^{\frac{1}{p}}\left[\zeta_2\left(x + \frac{x_0}{2}\right)\right]}e^{i\phi_2} \tag{8.32}$$

For non-Hamiltonian perturbations, the choice $\eta_1 = \eta_2 = 1$ and $\phi_1 = \phi_2 = 0$ gives

$$q(x,0) = \frac{1}{\cosh^{\frac{1}{p}}\left[\zeta\left(x - \frac{x_0}{2}\right)\right]} + \frac{1}{\cosh^{\frac{1}{p}}\left[\zeta\left(x + \frac{x_0}{2}\right)\right]} \qquad (8.33)$$

where

$$\zeta = \sqrt{\frac{2p^2}{1+p}} \qquad (8.34)$$

which represents an in-phase injection of pulses with equal amplitudes.

For Hamiltonian-type perturbations, the choice $\eta_1 = \eta_0$, $\eta_2 = 1$, and $\phi_1 = \phi_2 = 0$ gives

$$q(x,0) = \frac{\eta_0}{\cosh^{\frac{1}{p}}\left[\zeta_0\left(x - \frac{x_0}{2}\right)\right]} + \frac{1}{\cosh^{\frac{1}{p}}\left[\zeta\left(x + \frac{x_0}{2}\right)\right]} \qquad (8.35)$$

where

$$\zeta_0 = \eta_0^p \left(\frac{2p^2}{1+p}\right)^{\frac{1}{2}} \qquad (8.36)$$

which represents in-phase injection of solitons with unequal amplitudes.

8.2.3   Parabolic Law

For parabolic law nonlinearity, $F(s) = s + vs^2$, so that the NLSE is

$$iq_t + \frac{1}{2}q_{xx} + (|q|^2 + v|q|^4)q = i\epsilon R[q, q^*] \qquad (8.37)$$

Equation (8.37) is not integrable by the IST for $\epsilon = 0$, as discussed in chapter 5. However, (8.37) for $\epsilon = 0$ supports solitons of the form

$$q(x,t) = \frac{\eta}{[1 + a\cosh\{\zeta(x - vt - x_0)\}]^{\frac{1}{2}}} e^{i(-\kappa x + \omega t + \sigma_0)} \qquad (8.38)$$

where

$$\zeta \equiv \chi(\eta) = \eta\sqrt{2} \qquad (8.39)$$

$$\omega \equiv \psi(\eta, \kappa) = \frac{\eta^2}{4} - \frac{\kappa^2}{2} \qquad (8.40)$$

and

$$a = \sqrt{1 + \frac{4}{3}v\eta^2} \qquad (8.41)$$

In this chapter, it will be assumed that $v > 0$, although v could be negative as well, as seen in chapter 5. In fact, recall from chapter 5 that for $\epsilon = 0$, solitons exist for $v \in (-3/4A^2, \infty)$. Now, the 2-soliton solution of the parabolic law takes the asymptotic form

$$q(x, t) = \sum_{l=1}^{2} \frac{\eta_l}{[1 + a_l \cosh \{\zeta_l (x - vt - x_l)\}]^{\frac{1}{2}}} e^{i\left(-\kappa_l x + \omega_l t + \sigma_{0_l}\right)} \tag{8.42}$$

with

$$\zeta_l \equiv \chi(\eta_l) = \eta_l \sqrt{2} \tag{8.43}$$

$$\omega_l \equiv \psi(\eta_l, \kappa_l) = \frac{\eta_l^2 - 2\kappa_l^2}{4} \tag{8.44}$$

and

$$a_l = \sqrt{1 + \frac{4}{3} v \eta_l^2} \tag{8.45}$$

In the study of SSI with parabolic law nonlinearity, the initial pulse waveform is taken to be of the form

$$q(x, 0) = \frac{\eta_1}{\left[1 + a_1 \cosh \left\{\zeta_1 \left(x - \frac{x_0}{2}\right)\right\}\right]^{\frac{1}{2}}} e^{i\phi_1} + \frac{\eta_2}{\left[1 + a_2 \cosh \left\{\zeta_2 \left(x + \frac{x_0}{2}\right)\right\}\right]^{\frac{1}{2}}} e^{i\phi_2} \tag{8.46}$$

For non-Hamiltonian perturbations, the choice $\eta_1 = \eta_2 = 1$ and $\phi_1 = \phi_2 = 0$ gives

$$q(x, 0) = \frac{1}{\left[1 + \sqrt{1 + \frac{4}{3}v} \cosh \left\{\sqrt{2}\left(x - \frac{x_0}{2}\right)\right\}\right]^{\frac{1}{2}}}$$

$$+ \frac{1}{\left[1 + \sqrt{1 + \frac{4}{3}v} \cosh \left\{\sqrt{2}\left(x + \frac{x_0}{2}\right)\right\}\right]^{\frac{1}{2}}} \tag{8.47}$$

For Hamiltonian-type perturbations, the choice $\eta_1 = \eta_0$, $\eta_2 = 1$, and $\phi_1 = \phi_2 = 0$ gives

$$q(x, 0) = \frac{\eta_0}{\left[1 + \sqrt{1 + \frac{4}{3}v\eta_0} \cosh \left\{\eta_0\sqrt{2}\left(x - \frac{x_0}{2}\right)\right\}\right]^{\frac{1}{2}}}$$

$$+ \frac{1}{\left[1 + \sqrt{1 + \frac{4}{3}v} \cosh \left\{\sqrt{2}\left(x + \frac{x_0}{2}\right)\right\}\right]^{\frac{1}{2}}} \tag{8.48}$$

which represents an in-phase injection of pulses with unequal amplitudes.

8.2.4 Dual-Power Law

For dual-power law nonlinearity, $F(s) = s^p + vs^{2p}$ where $v < 0$, so that the NLSE is

$$iq_t + \frac{1}{2}q_{tt} + \left(|q|^{2p} + v|q|^{4p}\right)q = i\epsilon R[q, q^*] \qquad (8.49)$$

Equation (8.49) for $\epsilon = 0$, although not integrable by the IST, supports solitary waves of the form

$$q(x, t) = \frac{\eta}{[1 + a\cosh\{\zeta(x - vt - x_0)\}]^{\frac{1}{2p}}}e^{i(-\kappa x + \omega t + \sigma_0)} \qquad (8.50)$$

where

$$\zeta \equiv \chi(\eta) = \eta^p \left(\frac{4p^2}{1+p}\right)^{\frac{1}{2p}} \qquad (8.51)$$

$$\omega \equiv \psi(\eta, \kappa) = \frac{\eta^{2p}}{2p+2} - \frac{\kappa^2}{2} \qquad (8.52)$$

and

$$a = \sqrt{1 + \frac{v\zeta^2}{2p^2}\frac{(1+p)^2}{1+2p}} \qquad (8.53)$$

For dual-power law nonlinearity, solitons exist for

$$-\frac{2p^2}{\zeta^2}\frac{1+2p}{(1+p)^2} < v < 0 \qquad (8.54)$$

In this case, the 2-soliton solution of the NLSE (1) takes the asymptotic form

$$q(x, t) = \sum_{l=1}^{2} \frac{\eta_l}{[1 + a_l\cosh\{\zeta_l(x - vt - x_l)\}]^{\frac{1}{2p}}}e^{i(-\kappa_l x + \omega_l t + \sigma_{0_l})} \qquad (8.55)$$

with

$$\zeta_l \equiv \chi(\eta_l) = \eta_l^p \left(\frac{4p^2}{1+p}\right)^{\frac{1}{2p}} \qquad (8.56)$$

$$\omega_l \equiv \psi(\eta_l, \kappa_l) = \frac{\eta_l^{2p}}{2p+2} - \frac{\kappa_l^2}{2} \qquad (8.57)$$

and

$$a_l = \sqrt{1 + \frac{v\zeta_l^2}{2p^2}\frac{(1+p)^2}{1+2p}} \qquad (8.58)$$

In the study of SSI with dual-power law nonlinearity, the initial pulse waveform is taken to be of the form

$$q(x, 0) = \frac{\eta_1}{\left[1 + a_1 \cosh\left\{\zeta_1 \left(x - \frac{x_0}{2}\right)\right\}\right]^{\frac{1}{2p}}} e^{i\phi_1}$$

$$+ \frac{\eta_2}{\left[1 + a_2 \cosh\left\{\zeta_2 \left(x + \frac{x_0}{2}\right)\right\}\right]^{\frac{1}{2p}}} e^{i\phi_2} \qquad (8.59)$$

For non-Hamiltonian perturbations, the choice $\eta_1 = \eta_2 = 1$ and $\phi_1 = \phi_2 = 0$, gives

$$q(x, 0) = \frac{1}{\left[1 + a_1 \cosh\left\{\zeta \left(x - \frac{x_0}{2}\right)\right\}\right]^{\frac{1}{2p}}} + \frac{1}{\left[1 + a_2 \cosh\left\{\zeta \left(x + \frac{x_0}{2}\right)\right\}\right]^{\frac{1}{2p}}} \qquad (8.60)$$

where ζ is given by (8.34). For Hamiltonian-type perturbations, the choice $\eta_1 = \eta_0$, $\eta_2 = 1$, and $\phi_1 = \phi_2 = 0$ gives

$$q(x, 0) = \frac{\eta_0}{\left[1 + a_1 \cosh\left\{\zeta_0 \left(x - \frac{x_0}{2}\right)\right\}\right]^{\frac{1}{2p}}} + \frac{1}{\left[1 + a_2 \cosh\left\{\zeta \left(x + \frac{x_0}{2}\right)\right\}\right]^{\frac{1}{2p}}} \qquad (8.61)$$

where

$$\zeta_0 = \eta_0^p \sqrt{\frac{4p^2}{1 + p}} \qquad (8.62)$$

which represents in-phase injection of pulses with unequal amplitudes.

8.3 Quasi-Particle Theory

The QPT dates back to the 1981 appearance of the paper by Karpman and Solov'ev [213]. The mathematical approach to SSI will be studied using the QPT. In this theory, solitons are treated as particles. If two pulses are separated and each is close to a soliton, they can be written as the linear superposition of two soliton-like pulses as [185]

$$q(x, t) = q_1(x, t) + q_2(x, t) \qquad (8.63)$$

with

$$q_l(x, t) = A_l g[D_l(x - x_l)]e^{-i\{B_l(x - x_l) - \delta_l\}} \qquad (8.64)$$

where

$$D_l = \chi(A_l) \qquad (8.65)$$

and $l = 1, 2$, while A_l, B_l, D_l, T_l, and δ_l are functions of t. Here, A_l, D_l, and B_l do not represent the amplitude, width, and the frequency of the full wave form. However, they approach the amplitude, width, and frequency, respectively, for large separation, namely if $\Delta x = x_1 - x_2 \to \infty$, then $A_l \to \eta_l$, $D_l \to \zeta_l$, and $B_l \to \kappa_l$. Since the waveform is assumed to remain in the form of two pulses, the method is called the *quasi-particle approach*.

First, the equations for A_l, B_l, T_l, and δ_l using the soliton perturbation theory (SPT) will be derived. Substituting (8.63) into (8.1) yields

$$i\frac{\partial q_l}{\partial t} + \frac{1}{2}\frac{\partial^2 q_l}{\partial x^2} = i\epsilon R[q_l, q_l^*] - F(|q_l + q_{\bar{l}}|^2)|q_l + q_{\bar{l}}| \tag{8.66}$$

where $l = 1, 2$ and $\bar{l} = 3 - l$. By the SPT, the evolution equations are

$$\frac{dA_l}{dt} = F_1^{(l)}(A, \Delta x, \Delta\phi) + \epsilon M_l \tag{8.67}$$

$$\frac{dB_l}{dt} = F_2^{(l)}(A, \Delta x, \Delta\phi) + \epsilon N_l \tag{8.68}$$

$$\frac{dx_l}{dt} = -B_l - F_3(A, \Delta x, \Delta\phi) + \epsilon Q_l \tag{8.69}$$

$$\frac{d\delta_l}{dt} = \psi(A_l, B_l) + F_4(A, \Delta x, \Delta\phi) + \epsilon P_l \tag{8.70}$$

where

$$M_l = h_1(A_l) \int_{-\infty}^{\infty} \Re\{\hat{R}[q_l, q_l^*]e^{-i\phi_l}\}g(\tau_l)d\tau_l \tag{8.71}$$

$$N_l = h_2(A_l) \int_{-\infty}^{\infty} \Im\{\hat{R}[q_l, q_l^*]e^{-i\phi_l}\}g'(\tau_l)d\tau_l \tag{8.72}$$

$$Q_l = h_3(A_l) \int_{-\infty}^{\infty} \Re\{\hat{R}[q_l, q_l^*]e^{-i\phi_l}\}\tau_l g(\tau_l)d\tau_l \tag{8.73}$$

$$P_l = h_4(A_l) \int_{-\infty}^{\infty} \Im\{\hat{R}[q_l, q_l^*]e^{-i\phi_l}\}[g(\tau_l) - \tau_l g'(\tau_l)]d\tau_l \tag{8.74}$$

and the functions $F_1^{(l)}$, $F_2^{(l)}$, F_3, and F_4 evolve on using the SPT in (8.66), with the right side being treated as perturbation terms. The exact form of these functions can be obtained when a specific law of nonlinearity is considered. In (8.71)–(8.74), $h_j(A_l)$ for $1 \le j \le 4$ are by virtue of (8.65) and the type of nonlinearity that is considered. Also, in (8.71)–(8.74), $\Re$ and $\Im$ stand for the real and imaginary parts, respectively. Moreover, the following notations are used [185]

$$\hat{R}[q_l, q_l^*] = R[q_l, q_l^*] - F(|q_l + q_{\bar{l}}|^2)|q_l + q_{\bar{l}}| \tag{8.75}$$

$$\tau_l = D_l(x - x_l) \tag{8.76}$$

which represent the argument of the functional form of the lth soliton for various laws of nonlinearity, while the phase of the lth soliton in a fiber is given by

$$\phi_l = B_l(x - x_l) - \delta_l \tag{8.77}$$

The difference between the phases of the two solitons, namely $\phi_1 - \phi_2$, is given by

$$\Delta\phi = B\Delta x + \Delta\delta \tag{8.78}$$

while the separation of the solitons that is given by the distance between the centers of the solitons is represented as

$$\Delta x = x_1 - x_2 \tag{8.79}$$

$$\Delta\delta = \delta_1 - \delta_2 \tag{8.80}$$

The mean values of the amplitude and width of the two solitons are respectively defined as

$$A = \frac{1}{2}(A_1 + A_2) \tag{8.81}$$

and

$$B = \frac{1}{2}(B_1 + B_2) \tag{8.82}$$

Now, the difference between the amplitudes and widths of the two solitons are respectively given as

$$\Delta A = A_1 - A_2 \tag{8.83}$$

and

$$\Delta B = B_1 - B_2 \tag{8.84}$$

Finally, it is assumed that

$$|\Delta A| \ll A \tag{8.85}$$

$$|\Delta B| \ll 1 \tag{8.86}$$

$$|\Delta D| \ll D \tag{8.87}$$

$$A\Delta x \gg 1 \tag{8.88}$$

$$D\Delta x \gg 1 \tag{8.89}$$

$$|\Delta A|\Delta x \ll 1 \tag{8.90}$$

and

$$|\Delta D|\Delta x \ll 1 \tag{8.91}$$

From (8.67) to (8.70), one can now obtain

$$\frac{dA}{dt} = \epsilon M \tag{8.92}$$

$$\frac{dB}{dt} = \epsilon N \tag{8.93}$$

$$\frac{d(\Delta A)}{dt} = F_1^{(1)}(A, \Delta t, \Delta\phi) - F_1^{(2)}(A, \Delta x, \Delta\phi) + \epsilon \Delta M \tag{8.94}$$

$$\frac{d(\Delta B)}{dt} = F_2^{(1)}(A, \Delta x, \Delta\phi) - F_2^{(2)}(A, \Delta x, \Delta\phi) + \epsilon \Delta N \tag{8.95}$$

$$\frac{d(\Delta T)}{dt} = -\Delta B + \epsilon \Delta Q \tag{8.96}$$

$$\frac{d(\Delta\phi)}{dt} = \psi(A_1, B_1) - \psi(A_2, B_2) - B\Delta B$$
$$+ \frac{\Delta x}{2}\left(F_2^{(1)} + F_2^{(2)}\right) + \epsilon \Delta P + \epsilon B \Delta Q \tag{8.97}$$

where

$$M = \frac{1}{2}(M_1 + M_2) \tag{8.98}$$

$$N = \frac{1}{2}(N_1 + N_2) \tag{8.99}$$

and ΔM, ΔN, ΔQ, and ΔP are the variations of M, N, Q, and P, which are written as, for example

$$\Delta M = \frac{\partial M}{\partial A}\Delta A + \frac{\partial M}{\partial B}\Delta B \tag{8.100}$$

assuming that they are functions of A and B only, which is, in fact, true for most of the cases of interest; otherwise, the equations for

$$x = \frac{1}{2}(x_1 + x_2) \tag{8.101}$$

and

$$\phi = \frac{1}{2}(\phi_1 + \phi_2) \tag{8.102}$$

would have been necessary. The results derived in this section will now be utilized to show that SSI can indeed be suppressed in the presence of the perturbation terms given by (8.2) for the four cases of nonlinearity. In all four types of nonlinearity, corresponding to the initial waveform (8.13) are $A = 1$, $B = 0$, $\Delta A_0 = 0$, $\Delta B_0 = 0$, $\Delta T_0 = T_0$, and $\Delta\phi_0 = 0$. The four laws of nonlinearity are studied in the following subsections.

8.3.1   Kerr Law

In this case, (8.64) reduces to

$$q_l(x, t) = \frac{A_l}{\cosh[D_l(x - x_l)]} e^{-i B_l (x-x_l) + i \delta_l} \tag{8.103}$$

where

$$D_l \equiv \chi(A_l) = A_l \tag{8.104}$$

so that (8.66) transforms to

$$i \frac{\partial q_l}{\partial t} + \frac{1}{2} \frac{\partial^2 q_l}{\partial x^2} + |q_l|^2 q_l = i \epsilon R[q_l, q_l^*] - (q_l^2 q_{\bar l}^* + 2|q_l|^2 q_{\bar l}) \tag{8.105}$$

where $l = 1, 2$ and $\bar l = 3 - l$, and the separation

$$|q|^2 q = (|q_1|^2 q_1 + q_1^2 q_2^* + 2|q_1|^2 q_2) + (|q_2|^2 q_2 + q_2^2 q_1^* + 2|q_2|^2 q_1) \tag{8.106}$$

was used based on the degree of overlapping. By the SPT, the evolution equations are

$$\frac{d A_l}{dt} = (-1)^{l+1} 4 A^3 e^{-A \Delta x} \sin(\Delta \phi) + \epsilon M_l \tag{8.107}$$

$$\frac{d B_l}{dx} = (-1)^{l+1} 4 A^3 e^{-A \Delta x} \cos(\Delta \phi) + \epsilon N_l \tag{8.108}$$

$$\frac{d x_l}{dt} = -B_l - 2 A e^{-A \Delta x} \sin(\Delta \phi) + \epsilon Q_l \tag{8.109}$$

and

$$\frac{d \delta_l}{dt} = \frac{1}{2}(A_l^2 + B_l^2) - 2 A B e^{-A \Delta x} \sin(\Delta \phi) + 6 A^2 e^{-A \Delta x} \cos(\Delta \phi) + \epsilon P_l \tag{8.110}$$

where

$$M_l = \int_{-\infty}^{\infty} \Re\{\hat R[q_l, q_l^*] e^{-i \phi_l}\} \frac{1}{\cosh \tau_l} d\tau_l \tag{8.111}$$

$$N_l = -\int_{-\infty}^{\infty} \Im\{\hat R[q_l, q_l^*] e^{-i \phi_l}\} \frac{\tanh \tau_l}{\cosh \tau_l} d\tau_l \tag{8.112}$$

$$Q_l = \frac{1}{A_l^2} \int_{-\infty}^{\infty} \Re\{\hat R[q_l, q_l^*] e^{-i \phi_l}\} \frac{\tau_l}{\cosh \tau_l} d\tau_l \tag{8.113}$$

$$P_l = \frac{1}{A_l} \int_{-\infty}^{\infty} \Im\{\hat R[q_l, q_l^*] e^{-i \phi_l}\} \frac{(1 - \tau_l \tanh \tau_l)}{\cosh \tau_l} d\tau_l \tag{8.114}$$

and

$$\hat R[q_l, q_l^*] = R[q_l, q_l^*] - i(q_l^2 q_{\bar l}^* + 2|q_l|^2 q_{\bar l}^*) \tag{8.115}$$

The study for the Kerr law case will now be split into the following two subsections.

8.3.1.1 Non-Hamiltonian Perturbations

In the presence of non-Hamiltonian perturbation terms, as discussed after (8.2), the dynamical system of the soliton parameters by virtue of the SPT are

$$\frac{dA}{dt} = \epsilon\delta\frac{\Gamma\left(\frac{1}{2}\right)\Gamma(m+1)}{\Gamma\left(\frac{2m+3}{2}\right)}A^{2m+1} + 2\epsilon\sigma A^2 - \frac{2}{3}\epsilon\beta A(A^2 + 3B^2) \quad (8.116)$$

$$\frac{dB}{dt} = -\frac{4}{3}\epsilon\beta A^2 B \quad (8.117)$$

so that by virtue of (8.83), (8.84), (8.79), and (8.80)

$$\frac{d(\Delta A)}{dt} = 8A^3 e^{-A\Delta x}\sin(\Delta\phi)$$

$$+ \frac{\epsilon\delta}{2^{2m}}\frac{\Gamma\left(\frac{1}{2}\right)\Gamma\left(m+1\right)}{\Gamma\left(\frac{2m+3}{2}\right)}\sum_{r=0}^{2m+1}\binom{2m+1}{2r+1}(2A)^{2r+1}(\Delta A)^{2m-2r}$$

$$+ 4\epsilon\sigma A\Delta A - 2\epsilon\beta\left(A^2 + B^2\right)\Delta A - 4\epsilon\beta AB\Delta B \quad (8.118)$$

$$\frac{d(\Delta B)}{dt} = 8A^3 e^{-A\Delta x}\cos(\Delta\phi) - \frac{8}{3}\epsilon\beta AB\Delta A - \frac{4}{3}\epsilon\beta A^2\Delta B \quad (8.119)$$

$$\frac{d(\Delta x)}{dt} = -\Delta B + \epsilon\sigma \quad (8.120)$$

$$\frac{d(\Delta\phi)}{dt} = A\Delta A - \frac{4}{3}\epsilon\beta A^2 B\Delta x \quad (8.121)$$

where in (8.118)

$$\binom{n}{r} = \frac{n(n-1)\dots(n-r+1)}{r(r-1)\dots 3.2.1} \quad (8.122)$$

For the fixed point of the dynamical system, given by (8.116) and (8.117), with $A = 1$ and $B = 0$, one recovers

$$\beta = 3\sigma + \frac{3\delta}{2}\frac{\Gamma\left(\frac{1}{2}\right)\Gamma\left(m+1\right)}{\Gamma\left(\frac{2m+3}{2}\right)} \quad (8.123)$$

From (8.120) and (8.121), one has the coupled system of equations for $\Delta\phi$, the phase difference, Δx, and the soliton separation, with the fixed point $A = 1$ and $B = 0$ as follows

$$\frac{d^2(\Delta x)}{dt^2} + \frac{4}{3}\epsilon\beta\frac{d(\Delta x)}{dt} + 8e^{-\Delta x}\cos(\Delta\phi) = 0 \quad (8.124)$$

$$\frac{d^2(\Delta\phi)}{dt^2} + 2\epsilon(\beta - 2\sigma)\frac{d(\Delta\phi)}{dt}$$

$$- \frac{\epsilon \delta}{2^{2m}} \frac{\Gamma \left(\frac{1}{2}\right) \Gamma(m+1)}{\Gamma \left(\frac{2m+3}{2}\right)} \sum_{r=0}^{2m+1} \binom{2m+1}{2r+1} \left(\frac{d(\Delta\phi)}{dt}\right)^{2m-2r}$$

$$- 8e^{-\Delta x} \sin(\Delta\phi) = 0 \tag{8.125}$$

where in (8.124) and (8.125) β is given by (8.123). Equations (8.124) and (8.125) show that inserting filters produces a damping in both pulse separation and phase difference, as seen in Figures 8.1(a)–8.1(c).

8.3.1.2 Hamiltonian Perturbations

In the case of Hamiltonian perturbations, the dynamical system of the soliton parameters is

$$\frac{dA}{dt} = 0 \tag{8.126}$$

$$\frac{dB}{dt} = 0 \tag{8.127}$$

and

$$\frac{dx_0}{dt} = -B - 3\epsilon\gamma B^2 - \frac{\epsilon}{3}\{A^2(3\lambda + 2\mu + 3\gamma) + 3\alpha\} \tag{8.128}$$

so that by virtue of (8.79), (8.80), (8.83), and (8.84)

$$\frac{d(\Delta A)}{dt} = 8A^3 e^{-A\Delta x} \sin(\Delta\phi) \tag{8.129}$$

$$\frac{d(\Delta B)}{dt} = 8A^3 e^{-A\Delta x} \cos(\Delta\phi) \tag{8.130}$$

$$\frac{d(\Delta x)}{dt} = -\Delta B - \frac{3}{2}\epsilon\gamma B \Delta B - \frac{\epsilon}{6}(3\lambda + 2\mu + 3\gamma) A \Delta A \tag{8.131}$$

and

$$\frac{d(\Delta\phi)}{dt} = A\Delta A \tag{8.132}$$

Now

$$A = \frac{1}{2}(A_0 + 1) \tag{8.133}$$

$$B = 0 \tag{8.134}$$

$$\Delta A_0 = A_0 - 1 \tag{8.135}$$

$$\Delta B_0 = 0 \tag{8.136}$$

$$\Delta T_0 = T_0 \tag{8.137}$$

$$\Delta\phi_0 = 0 \tag{8.138}$$

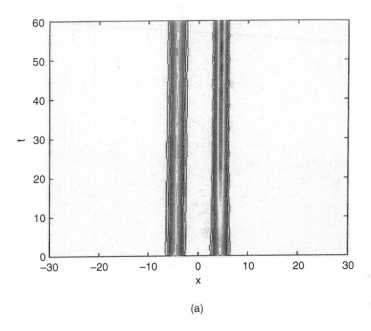

(a)

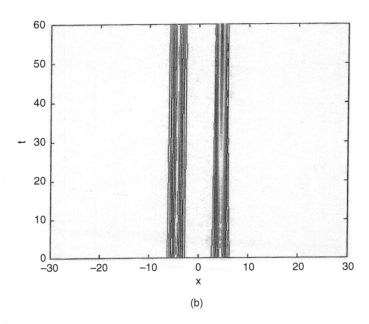

(b)

FIGURE 8.1
(a) SSI for $m = 0$, $\sigma = \delta = 0.005$, (b) SSI for $m = 1$, $\sigma = \delta = 0.005$.

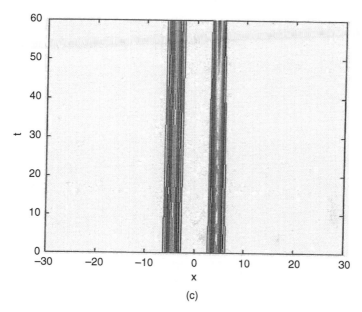

(c)

FIGURE 8.1 (Continued)
(c) SSI for $m = 2$, $\sigma = \delta = 0.005$.

and

$$\Delta\phi = \Delta\delta \tag{8.139}$$

so that

$$\frac{d(\Delta x)}{dt} + \frac{\epsilon}{6}(3\lambda + 2\mu + 3\gamma)\frac{d(\Delta\phi)}{dt} = -\Delta B \tag{8.140}$$

For $\Delta B = 0$

$$\Delta x = x_0 - \frac{\epsilon}{6}(3\lambda + 2\mu + 3\gamma)\Delta\delta \tag{8.141}$$

Since, $x_0 \sim O(1)$, $\Delta x \nrightarrow 0$ and thus the pulses do not collide during the transmission. This is observed in the numerical simulation in Figure 8.2.

8.3.2 Power Law

In this case, (8.64) is

$$q_l(x, t) = \frac{A_l}{\cosh^{\frac{1}{p}}[D_l(x - x_l)]}e^{-iB_l(x-x_l)+i\delta_l} \tag{8.142}$$

where

$$D_l \equiv \chi(A_l) = A_l^p\left(\frac{2p^2}{1+p}\right)^{\frac{1}{2}} \tag{8.143}$$

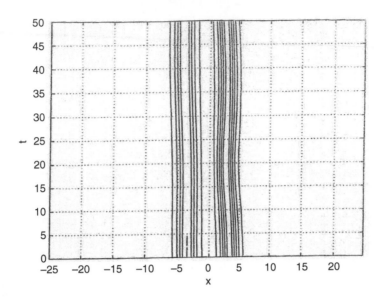

FIGURE 8.2
SSI with $\gamma = 0.14$.

so that using (8.63) in (8.1) yields

$$i\frac{\partial q_l}{\partial t} + \frac{1}{2}\frac{\partial^2 q_l}{\partial x^2} = i\epsilon R[q_l, q_l^*] - \left[\sum_{r=0}^{p}\binom{p}{r}q_1^{p-r}q_2^r\right]$$

$$\left[\sum_{r=0}^{p}\binom{p}{r}(q_1^*)^{p-r}(q_2^*)^r\right](q_1 + q_2) \qquad (8.144)$$

Here, the separation

$$|q|^{2p}q = \left[\sum_{r=0}^{p}\binom{p}{r}q_1^{p-r}q_2^r\right]\left[\sum_{r=0}^{p}\binom{p}{r}(q_1^*)^{p-r}(q_2^*)^r\right](q_1 + q_2) \qquad (8.145)$$

was used based on the degree of overlapping. By the SPT, the evolution equations are

$$\frac{dA_l}{dt} = F_1^{(l)}(A, \Delta x, \Delta\phi; p) + \epsilon M_l \qquad (8.146)$$

$$\frac{dB_l}{dt} = F_2^{(l)}(A, \Delta x, \Delta\phi; p) + \epsilon N_l \qquad (8.147)$$

$$\frac{dT_l}{dt} = -B_l - F_3(A, \Delta x, \Delta\phi; p) + \epsilon Q_l \qquad (8.148)$$

and

$$\frac{d\delta_l}{dt} = \frac{B_l^2}{2} + \frac{A_l^{2p}}{p+1} + F_4(A, \Delta x, \Delta\phi; p) + \epsilon P_l \qquad (8.149)$$

where, for power law

$$M_l = \frac{1}{2-p}\left(\frac{2p^2}{1+p}\right)^{\frac{p-3}{2p}} \frac{\Gamma\left(\frac{1}{p}+\frac{1}{2}\right)}{\Gamma\left(\frac{1}{p}\right)\Gamma\left(\frac{1}{2}\right)} \int_{-\infty}^{\infty} \Re\left\{\hat{R}[q_l,q_l^*]e^{-i\phi_l}\right\} \frac{1}{\cosh^{\frac{1}{p}}\tau_l}d\tau_l$$

(8.150)

$$N_l = \frac{2}{p}A_l^{p-1}\left(\frac{2p^2}{1+p}\right)^{\frac{p-1}{2p}} \frac{\Gamma\left(\frac{1}{p}+\frac{1}{2}\right)}{\Gamma\left(\frac{1}{p}\right)\Gamma\left(\frac{1}{2}\right)} \int_{-\infty}^{\infty} \Im\left\{\hat{R}[q_l,q_l^*]e^{-i\phi_l}\right\} \frac{\tanh\tau_l}{\cosh^{\frac{1}{p}}\tau_l}d\tau_l$$

(8.151)

$$Q_l = \frac{1}{A_l^{p+1}}\left(\frac{p+1}{2p^2}\right)^{\frac{p+2}{2p}} \frac{\Gamma\left(\frac{1}{p}+\frac{1}{2}\right)}{\Gamma\left(\frac{1}{p}\right)\Gamma\left(\frac{1}{2}\right)} \int_{-\infty}^{\infty} \Re\left\{\hat{R}[q_l,q_l^*]e^{-i\phi_l}\right\} \frac{\tau_l}{\cosh^{\frac{1}{p}}\tau_l}d\tau_l$$

(8.152)

and

$$P_l = \frac{1}{A_l^{p+1}}\left(\frac{2p^2}{p+1}\right)^{\frac{p+1}{2p}} \frac{\Gamma\left(\frac{1}{p}+\frac{1}{2}\right)}{\Gamma\left(\frac{1}{p}\right)\Gamma\left(\frac{1}{2}\right)}$$
$$\cdot \int_{-\infty}^{\infty} \Im\left\{\hat{R}[q_l,q_l^*]e^{-i\phi_l}\right\} \frac{(1-\tau_l\tanh\tau_l)}{\cosh^{\frac{1}{p}}\tau_l}d\tau_l$$

(8.153)

In addition, the following notations are used

$$\hat{R}[q_l,q_l^*] = R[q_l,q_l^*] - i\left[\sum_{r=0}^{p}\binom{p}{r}q_1^{p-r}q_2^r\right]\left[\sum_{r=0}^{p}\binom{p}{r}(q_1^*)^{p-r}(q_2^*)^r\right]$$
$$(q_1+q_2)+i|q_1|^{2p}q_1$$

(8.154)

For power law, the study will now be split into the following two cases.

8.3.2.1 Non-Hamiltonian Perturbations

In the presence of non-Hamiltonian perturbation terms, as mentioned after (8.2), the dynamical system of the soliton parameters by virtue of the SPT are

$$\frac{dA}{dt} = \frac{2\epsilon\delta}{2-p}\left(\frac{1+p}{2p^2}\right)^{\frac{1}{2p}} \frac{\Gamma\left(\frac{1}{p}+\frac{1}{2}\right)}{\Gamma\left(\frac{1}{p}\right)} \frac{\Gamma\left(\frac{m+1}{p}\right)}{\Gamma\left(\frac{m+1}{p}+\frac{1}{2}\right)}A^{2m+1}$$

$$+ \frac{\epsilon\sigma}{2-p}\left(\frac{1+p}{2p^2}\right)^{\frac{2p}{p+1}} \frac{\Gamma\left(\frac{1}{p}+\frac{1}{2}\right)}{\Gamma\left(\frac{1}{p}\right)\Gamma\left(\frac{1}{2}\right)} \int_{-\infty}^{\infty} \frac{1}{\cosh^{\frac{2}{p}}\tau}\left(\int_{-\infty}^{\tau}\frac{ds}{\cosh^{\frac{2}{p}}s}\right)d\tau A^{3-p}$$

$$- \frac{2\epsilon\beta}{p^2(2-p)}\left(\frac{2p^2}{p+1}\right)^{\frac{2p-1}{2p}} \frac{\Gamma\left(\frac{1}{p}+\frac{1}{2}\right)}{\Gamma\left(\frac{1}{p}\right)} \frac{\Gamma\left(\frac{p+1}{p}\right)}{\Gamma\left(\frac{p+1}{p}+\frac{1}{2}\right)}A^{2p+1}$$

$$- \frac{2\epsilon\beta}{2-p}\left(\frac{2p^2}{p+1}\right)^{\frac{1}{2p}}AB^2 - \frac{2\epsilon\beta}{p^2(2-p)}\left(\frac{2p^2}{p+1}\right)^{\frac{2p-1}{2p}}A^{2p+1}$$

(8.155)

and

$$\frac{dB}{dt} = \frac{4\epsilon\beta}{p^2}\left(\frac{2p^2}{p+1}\right)^{\frac{3p-2}{2p}}\left(\frac{p-2}{p+2}\right)BA^{2p} \tag{8.156}$$

so that by virtue of (8.79), (8.80), (8.83), and (8.84)

$$\frac{d(\Delta A)}{dt} = F_1^{(1)}(A, \Delta x, \Delta\phi) - F_1^{(2)}(A, \Delta x, \Delta\phi) + \frac{\epsilon\delta}{2-p}\left(\frac{1+p}{2p^2}\right)^{\frac{1}{2p}}$$

$$\cdot\frac{\Gamma\left(\frac{1}{p}+\frac{1}{2}\right)}{\Gamma\left(\frac{1}{p}\right)}\frac{\Gamma\left(\frac{m+1}{p}\right)}{\Gamma\left(\frac{m+1}{p}+\frac{1}{2}\right)}\sum_{r=0}^{2m+1}\binom{2m+1}{2r+1}(2A)^{2r+1}(\Delta A)^{2m-2r}$$

$$+\frac{\epsilon\sigma}{2-p}\left(\frac{1+p}{2p^2}\right)^{\frac{2p}{p+1}}\frac{\Gamma\left(\frac{1}{p}+\frac{1}{2}\right)}{\Gamma\left(\frac{1}{p}\right)\Gamma\left(\frac{1}{2}\right)}$$

$$\cdot\int_{-\infty}^{\infty}\frac{1}{\cosh^{\frac{2}{p}}\tau}\left(\int_{-\infty}^{\tau}\frac{ds}{\cosh^{\frac{2}{p}}s}\right)d\tau$$

$$\cdot\sum_{r=0}^{2m+1}\binom{3-p}{2r+1}(2A)^{2-p+2r}(\Delta A)^{2r+1} + \frac{\epsilon\beta}{p^2(2-p)}\left(\frac{2p^2}{p+1}\right)^{\frac{2p-1}{2p}}\frac{\Gamma\left(\frac{1}{p}+\frac{1}{2}\right)}{}$$

$$\cdot\Gamma\left(\frac{1}{p}\right)\frac{\Gamma\left(\frac{p+1}{p}\right)}{\Gamma\left(\frac{p+1}{p}+\frac{1}{2}\right)}\sum_{r=0}^{2m+1}\binom{2m+1}{2r+1}(2A)^{2r+1}(\Delta A)^{2m-2r}$$

$$-\frac{2\epsilon\beta}{2-p}\left(\frac{2p^2}{p+1}\right)^{\frac{1}{2p}}AB\Delta B - \frac{\epsilon\beta}{p^2(2-p)}\left(\frac{2p^2}{p+1}\right)^{\frac{2p-1}{2p}}$$

$$\cdot\sum_{r=0}^{2m+1}\binom{2m+1}{2r+1}(2A)^{2r+1}(\Delta A)^{2m-2r} \tag{8.157}$$

$$\frac{d(\Delta B)}{dt} = F_2^{(1)}(A, \Delta x, \Delta\phi) - F_2^{(2)}(A, \Delta x, \Delta\phi)$$

$$+\frac{4\epsilon\beta}{p^2}\left(\frac{2p^2}{p+1}\right)^{\frac{3p-2}{2p}}\left(\frac{p-2}{p+2}\right)[\Delta BA^{2p} + 2pBA^{2p+1}\Delta A] \tag{8.158}$$

$$\frac{d(\Delta x)}{dt} = -\Delta B + 2\epsilon\sigma\left(\frac{A_1^{p+2}}{D_1^3} - \frac{A_2^{p+2}}{D_2^3}\right)\left(\frac{2p^2}{1+p}\right)^{\frac{1}{2}}$$

$$\frac{\Gamma\left(\frac{1}{p}+\frac{1}{2}\right)}{\Gamma\left(\frac{1}{p}\right)\Gamma\left(\frac{1}{2}\right)}\int_{-\infty}^{\infty}\frac{\tau}{\cosh^{\frac{2}{p}}\tau}\left(\int_{-\infty}^{\tau}\frac{ds}{\cosh^{\frac{2}{p}}s}\right)d\tau \tag{8.159}$$

and

$$\frac{d(\Delta\phi)}{dt} = \frac{A_1^{2p} - A_2^{2p}}{p+1} - \frac{4\epsilon\beta}{p^2}\left(\frac{2p^2}{p+1}\right)^{\frac{3p-1}{2p}}\left(\frac{p-2}{p+2}\right)A^{2p}B\Delta x \qquad (8.160)$$

For the fixed point of the dynamical system given by (8.155) and (8.156), with $A = 1$ and $B = 0$, one gets

$$\beta = \frac{1}{2\Gamma\left(\frac{2}{p}\right) - \Gamma\left(\frac{1}{p} + \frac{1}{2}\right)}\left[\delta(p+1)\frac{\Gamma\left(\frac{2}{p}\right)\Gamma\left(\frac{1}{p} + \frac{1}{2}\right)}{\Gamma\left(\frac{1}{p}\right)}\frac{\Gamma\left(\frac{m+1}{p}\right)}{\Gamma\left(\frac{m+1}{p} + \frac{1}{2}\right)}\right.$$

$$\left. + \sigma p^2\left(\frac{p+1}{2p^2}\right)^{\frac{(3p-1)(2p+1)}{2p(p+1)}}\frac{\Gamma\left(\frac{2}{p}\right)\Gamma\left(\frac{1}{p} + \frac{1}{2}\right)}{\Gamma\left(\frac{1}{p}\right)\Gamma\left(\frac{1}{2}\right)}\int_{-\infty}^{\infty}\frac{1}{\cosh^{\frac{2}{p}}\tau}\left(\int_{-\infty}^{\tau}\frac{ds}{\cosh^{\frac{2}{p}}s}\right)d\tau\right]$$

$$(8.161)$$

Thus, by (8.159) and (8.160) one gets the coupled system of equations in Δx and $\Delta\phi$ for the fixed point with $A = 1$ and $B = 0$ as

$$\frac{d^2(\Delta x)}{dt^2} - \frac{4\epsilon\beta}{p^2}\left(\frac{2p^2}{p+1}\right)^{\frac{3p-2}{2p}}\left(\frac{p-2}{p+2}\right)\frac{d(\Delta x)}{dt} + F_2^{(1)} - F_2^{(1)} = 0 \qquad (8.162)$$

and

$$\frac{d^2(\Delta\phi)}{dt^2} = \frac{2p}{p+1}\left[A_1^{2p-1}\frac{dA_1}{dt} - A_2^{2p-1}\frac{dA_2}{dt}\right] \qquad (8.163)$$

where in (8.162) β is given by (8.161). From (8.162) one can observe that a damping is introduced in the soliton separation and the coefficient of the damping term is positive as $0 < p < 2$. In Figures 8.3(a) and 8.3(b), numerical simulations show that the suppression of the SSI is achieved for power law as proved in the QPT.

8.3.2.2 Hamiltonian Perturbations

In the presence of Hamiltonian perturbation terms, as given by (8.2), the dynamical system of the soliton parameters by virtue of the SPT are

$$\frac{dA}{dt} = 0 \qquad (8.164)$$

$$\frac{dB}{dt} = 0 \qquad (8.165)$$

and

$$\frac{dx_0}{dt} = -B - \frac{\epsilon}{2}A^2(3\lambda + 2\mu)\frac{\Gamma\left(\frac{1}{2} + \frac{1}{p}\right)\Gamma\left(\frac{2}{p}\right)}{\Gamma\left(\frac{1}{2} + \frac{2}{p}\right)\Gamma\left(\frac{1}{p}\right)}$$

$$- \epsilon\left(\mu + 3\gamma B^2\right) + \frac{3\epsilon\gamma D^3}{p^2}\left[\frac{\Gamma\left(\frac{p-1}{p}\right)\Gamma\left(\frac{1}{2} + \frac{1}{p}\right)}{\Gamma\left(\frac{1}{p}\right)\Gamma\left(\frac{3}{2} + \frac{2}{p}\right)} + 1\right] \qquad (8.166)$$

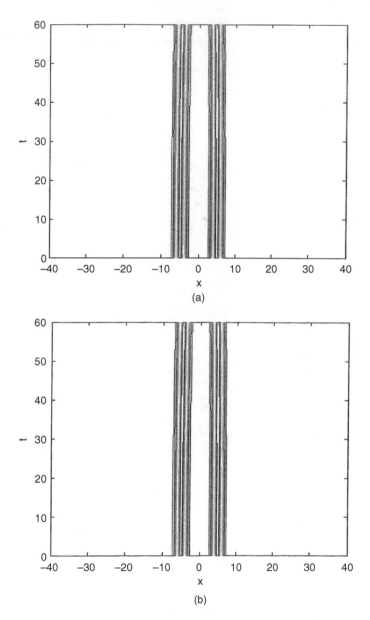

FIGURE 8.3
(a) SSI for $m = 1$, $p = 1/2$, $\delta = 0.001$, and (b) SSI for $m = 2$, $p = 1/2$, $\delta = 0.001$.

so that by virtue of (8.79), (8.80), (8.83), and (8.84)

$$\frac{d(\Delta A)}{dt} = F_1^{(1)}(A, \Delta x, \Delta\phi) - F_1^{(2)}(A, \Delta x, \Delta\phi) \qquad (8.167)$$

$$\frac{d(\Delta B)}{dt} = F_2^{(1)}(A, \Delta x, \Delta\phi) - F_2^{(2)}(A, \Delta x, \Delta\phi) \qquad (8.168)$$

$$\frac{d(\Delta x)}{dt} = -\Delta B - \frac{\epsilon}{4} A \Delta A (3\lambda + 2\mu) \frac{\Gamma(\frac{1}{2} + \frac{1}{p}) \Gamma(\frac{2}{p})}{\Gamma(\frac{1}{2} + \frac{2}{p}) \Gamma(\frac{1}{p})}$$

$$- \frac{3\epsilon\gamma}{4p^2} \Delta D [12D^2 + (\Delta D)^2]$$

$$+ \frac{3\epsilon\gamma}{p^2} \Delta D [12D^2 + (\Delta D)^2] \frac{\Gamma(\frac{p-1}{p}) \Gamma(\frac{1}{2} + \frac{1}{p})}{\Gamma(\frac{1}{p}) \Gamma(\frac{3}{2} + \frac{1}{p})} \qquad (8.169)$$

and

$$\frac{d(\Delta\phi)}{dt} = \frac{A_1^{2p} - A_2^{2p}}{p + 1} \qquad (8.170)$$

Now (8.169) can be rewritten as

$$\frac{d(\Delta x)}{dt} = -\Delta B - \frac{\epsilon}{4}(3\lambda + 2\mu)g_1 \left\{\frac{d(\Delta\phi)}{dt}\right\} \frac{\Gamma(\frac{1}{2} + \frac{1}{p}) \Gamma(\frac{2}{p})}{\Gamma(\frac{1}{2} + \frac{2}{p}) \Gamma(\frac{1}{p})}$$

$$- \frac{3\epsilon\gamma}{4p^2}g_2 \left\{\frac{d(\Delta\phi)}{dt}\right\} + \frac{3\epsilon\gamma}{p^2}g_2 \left\{\frac{d(\Delta\phi)}{dt}\right\} \frac{\Gamma(\frac{p-1}{p}) \Gamma(\frac{1}{2} + \frac{1}{p})}{\Gamma(\frac{1}{p}) \Gamma(\frac{3}{2} + \frac{1}{p})} \qquad (8.171)$$

For in-phase injection of solitons with unequal amplitudes

$$A = \frac{1}{2}(A_0 + 1) \qquad (8.172)$$

$$B = 0 \qquad (8.173)$$

$$\Delta A_0 = A_0 - 1 \qquad (8.174)$$

$$\Delta B_0 = 0 \qquad (8.175)$$

$$\Delta T_0 = T_0 \qquad (8.176)$$

$$\Delta\phi_0 = 0 \qquad (8.177)$$

and

$$\Delta\phi = \Delta\delta \qquad (8.178)$$

so that for $\Delta B = 0$

$$\Delta x = x_0 - \frac{\epsilon}{6}(3\lambda + 2\mu)$$

$$- \frac{\epsilon}{4}(3\lambda + 2\mu)h_1 \left\{\frac{d(\Delta\phi)}{dt}\right\} \frac{\Gamma(\frac{1}{2} + \frac{1}{p}) \Gamma(\frac{2}{p})}{\Gamma(\frac{1}{2} + \frac{2}{p}) \Gamma(\frac{1}{p})}$$

$$- \frac{3\epsilon\gamma}{4p^2}h_2 \left\{\frac{d(\Delta\phi)}{dt}\right\} + \frac{3\epsilon\gamma}{p^2}h_2 \left\{\frac{d(\Delta\phi)}{dt}\right\} \frac{\Gamma(\frac{p-1}{p}) \Gamma(\frac{1}{2} + \frac{1}{p})}{\Gamma(\frac{1}{p}) \Gamma(\frac{3}{2} - \frac{1}{p})} \qquad (8.179)$$

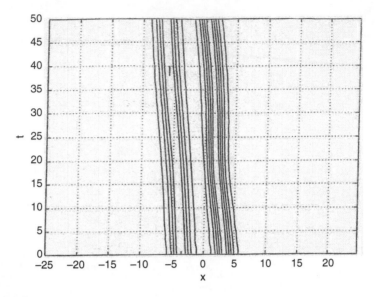

FIGURE 8.4
SSI with $p = 1/2$, $\gamma = 0.14$, $\lambda = \nu = 0.5$.

where

$$h_j(s) = \int g_j(s)ds \tag{8.180}$$

for $j = 1, 2$. Thus

$$\Delta x = x_0 + O(\epsilon) \tag{8.181}$$

Now, $x_0 \sim O(1)$ so that $\Delta x \nrightarrow 0$ and thus the pulses do not collide during the transmission. This can be observed in the numerical simulation in Figure 8.4.

8.3.3 Parabolic Law

For parabolic law nonlinearity

$$q_l(x, t) = \frac{A_l}{[1 + a_l \cosh\{D_l(x - x_l)\}]^{\frac{1}{2}}} e^{-i B_l(x - x_l) + i \delta_l} \tag{8.182}$$

where

$$D_l \equiv \chi(A_l) = A_l \sqrt{2} \tag{8.183}$$

Substituting (8.63) into (8.1) gives

$$i\frac{\partial q_l}{\partial t} + \frac{1}{2}\frac{\partial^2 q_l}{\partial x^2} + (|q_l|^2 + \nu|q_l|^4)q_l = i\epsilon R[q_l, q_l^*] - (q_l^2 q_l^* + 2|q_l|^2 q_l)$$
$$- \nu\left[q_l^3\left(q_l^*\right)^2 + 2|q_l|^2 q_l^2 q_l^* + 3|q_l|^4 q_l + 3|q_l|^2 q_l^* q_l^2 + 6|q_l|^2|q_l|^2 q_l\right] \tag{8.184}$$

where $l = 1, 2$ and $\bar{l} = 3 - l$. Here, the separation

$$|q|^2 q + v|q|^4 q = (|q_1|^2 q_1 + q_1^2 q_2^* + 2|q_1|^2 q_2) + (|q_2|^2 q_2 + q_2^2 q_1^* + 2|q_2|^2 q_1)$$
$$+ v[|q_1|^4 q_1 + q_1^3 (q_2^*)^2 + 2|q_1|^2 q_1^2 q_2^* + 3|q_1|^4 q_2$$
$$+ 3|q_1|^2 q_1^* q_2^2 + 6|q_1|^2 |q_2|^2 q_1]$$
$$+ v[|q_2|^4 q_2 + q_2^3 (q_1^*)^2 + 2|q_2|^2 q_2^2 q_1^* + 3|q_2|^4 q_1 + 3|q_2|^2 q_2^* q_1^2 + 6|q_1|^2 |q_2|^2 q_2]$$

$$(8.185)$$

was used based on the degree of overlapping. By virtue of the SPT, the evolution equations are

$$\frac{dA_l}{dt} = F_1^{(l)}(A, \Delta x, \Delta \phi; v) + \epsilon M_l \qquad (8.186)$$

$$\frac{dB_l}{dt} = F_2^{(l)}(A, \Delta x, \Delta \phi; v) + \epsilon N_l \qquad (8.187)$$

$$\frac{dT_l}{dt} = -B_l - F_3(A, \Delta x, \Delta \phi; v) + \epsilon Q_l \qquad (8.188)$$

and

$$\frac{d\delta_l}{dt} = \frac{A_l^2}{4} + \frac{B_l^2}{2} + F_4(A, \Delta x, \Delta \phi; v) + \epsilon P_l \qquad (8.189)$$

where

$$M_l = h_1^{(1)}(A_l) \int_{-\infty}^{\infty} \Re \left\{ \hat{R}[q_l, q_l^*] e^{-i\phi_l} \right\} \frac{d\tau_l}{(1 + a_l \cosh \tau_l)^{\frac{1}{2}}} d\tau_l \qquad (8.190)$$

$$N_l = h_2^{(1)}(A_l) \int_{-\infty}^{\infty} \Im \left\{ \hat{R}[q_l, q_l^*] e^{-i\phi_l} \right\} \frac{\sinh \tau_l}{(1 + a_l \cosh \tau_l)^{\frac{1}{2}}} d\tau_l \qquad (8.191)$$

$$Q_l = h_3^{(1)}(A_l) \int_{-\infty}^{\infty} \Re \left\{ \hat{R}[q_l, q_l^*] e^{-i\phi_l} \right\} \frac{\tau_l}{(1 + a_l \cosh \tau_l)^{\frac{1}{2}}} d\tau_l \qquad (8.192)$$

and

$$P_l = h_4^{(1)}(A_l) \int_{-\infty}^{\infty} \Im \left\{ \hat{R}[q_l, q_l^*] e^{-i\phi_l} \right\} \frac{(1 - a_l \tau_l \sinh \tau_l)}{(1 + a_l \cosh \tau_l)^{\frac{1}{2}}} d\tau_l \qquad (8.193)$$

Also, the following notations are used

$$\hat{R}[q_l, q_l^*] = R[q_l, q_l^*] - (q_l^2 q_{\bar{l}}^* + 2|q_l|^2 q_{\bar{l}}) - v[q_l^3 (q_{\bar{l}}^*)^2 + 2|q_l|^2 q_l^2 q_{\bar{l}}^* + 3|q_l|^4 q_{\bar{l}}$$
$$+ 3|q_l|^2 q_l^* q_{\bar{l}}^2 + 6|q_l|^2 |q_{\bar{l}}|^2 q_l] \qquad (8.194)$$

For the parabolic law case, the study will be split into the following two subsections.

8.3.3.1 Non-Hamiltonian Perturbations

In the presence of non-Hamiltonian perturbation terms, as given by (8.2), the dynamical system of the soliton parameters by virtue of the SPT is

$$
\frac{dA}{dt} = \frac{\epsilon \delta A^{2m+1}}{2^m a^{m-1}} F\left(m+1, m+1, m+\frac{3}{2}, \frac{a-1}{2a}\right) B\left(\frac{m+1}{p}, \frac{1}{2}\right)
$$

$$
+ \epsilon \sigma a^2 \frac{\sqrt{2}}{2} \int_{-\infty}^{\infty} \frac{1}{1 + a \cosh \tau} \left(\int_{-\infty}^{\tau} \frac{ds}{1 + a \cosh s}\right) d\tau
$$

$$
- \epsilon \beta a^2 A^3 \left[\frac{a^2}{(a^2-1)^{\frac{3}{2}}} \tan^{-1}\sqrt{\frac{a-1}{a+1}} - \frac{1}{2(a^2-1)}\right]
$$

$$
- 4\sqrt{2}\epsilon\beta \frac{a^2 B^2}{\sqrt{a^2-1}} \tan^{-1}\sqrt{\frac{a-1}{a+1}} \tag{8.195}
$$

and

$$
\frac{dB}{dt} = -\epsilon\beta \frac{\sqrt{2}A^3 B}{E} \left[\frac{a^2}{(a^2-1)^{\frac{3}{2}}} \tan^{-1}\sqrt{\frac{a-1}{a+1}} - \frac{1}{2(a^2-1)}\right] \tag{8.196}
$$

where E is the energy of the soliton given by

$$
E = \int_{-\infty}^{\infty} |q|^2 dx = \frac{2\sqrt{2}A}{\sqrt{a^2-1}} \tan^{-1}\sqrt{\frac{a-1}{a+1}} \tag{8.197}
$$

For the fixed point of the dynamical system, given by (8.195) and (8.196), with $A = 1$ and $B = 0$, one recovers

$$
\beta = \frac{\sigma\sqrt{2}}{2} \frac{\int_{-\infty}^{\infty} \frac{1}{1+a\cosh\tau}\left(\int_{-\infty}^{\tau} \frac{ds}{1+a\cosh s}\right)d\tau}{\frac{a^2}{(a^2-1)^{\frac{3}{2}}} \tan^{-1}\sqrt{\frac{a-1}{a+1}} - \frac{1}{2(a^2-1)}}
$$

$$
+ \frac{\delta}{2^m a^{m+1}} \frac{F\left(m+1, m+1, m+\frac{3}{2}; \frac{a-1}{2a}\right) B\left(\frac{m+1}{p}, \frac{1}{2}\right)}{\frac{a^2}{(a^2-1)^{\frac{3}{2}}} \tan^{-1}\sqrt{\frac{a-1}{a+1}} - \frac{1}{2(a^2-1)}} \tag{8.198}
$$

Thus, using (8.79)–(8.80), one can obtain

$$
\frac{d^2(\Delta x)}{dt^2} + \epsilon\beta G \frac{d(\Delta x)}{dt} + F_2^{(1)} - F_2^{(2)} = 0 \tag{8.199}
$$

where β is given by (8.198) and $G > 0$ represents the coefficient of $-\epsilon\beta\Delta B$ in $d(\Delta B)/dt = dB_1/dt - dB_2/dt$. Now, (8.199) shows a damping in the separation of solitons, thus proving that there will be a suppression of SSI in the presence of the perturbation terms given by (8.2). The numerical simulations in Figure 8.5 show that the suppression of SSI is achieved for parabolic law as proved in the QPT.

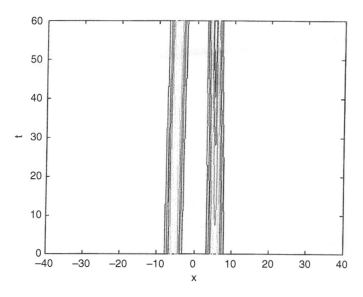

FIGURE 8.5
SSI with $m = 0$, $\delta = 0.001$, $\sigma = 0.001$.

8.3.3.2 Hamiltonian Perturbations

In the presence of Hamiltonian perturbation terms, as given by (8.2), the dynamical system of the soliton parameters by virtue of the SPT is

$$\frac{dA}{dt} = 0 \tag{8.200}$$

$$\frac{dB}{dt} = 0 \tag{8.201}$$

and

$$\frac{dx_0}{dt} = -B - \frac{\epsilon\alpha}{E}\frac{2\sqrt{2}A}{\sqrt{a^2-1}}\tan^{-1}\sqrt{\frac{a-1}{a+1}} - \frac{\epsilon\sqrt{2}}{2E}\frac{A^3}{(a^2-1)^{\frac{3}{2}}}(3\lambda + 2\mu)$$

$$\cdot\left[\sqrt{a^2-1} - 2\tan^{-1}\sqrt{\frac{a-1}{a+1}}\right] - 3\sqrt{2}\frac{\epsilon\gamma A}{E}\left[\left(\frac{a^4A^2}{2} + \frac{2\kappa^2}{\sqrt{a^2-1}}\right)\right.$$

$$\left.\cdot\tan^{-1}\sqrt{\frac{a-1}{a+1}} - \frac{a^2A^2}{4}\sqrt{a^2-1}\right] \tag{8.202}$$

From (8.79), (8.80), (8.83), and (8.84), one can now conclude that

$$\frac{d(\Delta A)}{dt} = F_1^{(1)}(A, \Delta x, \Delta\phi) - F_1^{(2)}(A, \Delta x, \Delta\phi) \tag{8.203}$$

$$\frac{d(\Delta B)}{dt} = F_2^{(1)}(A, \Delta x, \Delta\phi) - F_2^{(2)}(A, \Delta x, \Delta\phi) \tag{8.204}$$

$$\frac{d(\Delta x)}{dt} = \frac{dx_1}{dt} - \frac{dx_2}{dt} \tag{8.205}$$

and

$$\frac{d(\Delta\phi)}{dt} = \frac{1}{2}A\Delta A \qquad (8.206)$$

From (8.202), one can get

$$\frac{d(\Delta x)}{dt} = -\Delta B + \epsilon G\left(\alpha, \lambda, \mu, \gamma, \sigma\right) g\left\{\frac{d(\Delta\phi)}{dt}\right\} \qquad (8.207)$$

where G is the functional form that depends on the said parameters. For in-phase injection of solitons with unequal amplitudes

$$A = \frac{1}{2}(A_0 + 1) \qquad (8.208)$$

$$B = 0 \qquad (8.209)$$

$$\Delta A_0 = A_0 - 1 \qquad (8.210)$$

$$\Delta B_0 = 0 \qquad (8.211)$$

$$\Delta T_0 = T_0 \qquad (8.212)$$

$$\Delta\phi_0 = 0 \qquad (8.213)$$

and

$$\Delta\phi = \Delta\delta \qquad (8.214)$$

so that one can obtain from (8.207) for $\Delta B = 0$

$$\Delta x = x_0 + \epsilon G\left(\alpha, \lambda, \mu, \gamma, \sigma\right) h\left\{\frac{d(\Delta\phi)}{dt}\right\} \qquad (8.215)$$

where

$$h_j(s) = \int g_j(s)ds \qquad (8.216)$$

for $j = 1, 2$. Thus

$$\Delta x = x_0 + O(\epsilon) \qquad (8.217)$$

Now, $x_0 \sim O(1)$ so that $\Delta x \nrightarrow 0$ and thus the pulses do not collide during the transmission, as shown in the numerical simulations in Figure 8.6.

8.3.4 Dual-Power Law

For dual-power law nonlinearity

$$q_l(x, t) = \frac{A_l}{[1 + a_l \cosh\{D_l(x - x_l)\}]^{\frac{1}{2}}}e^{-iB_l(x-x_l)+i\delta_l} \qquad (8.218)$$

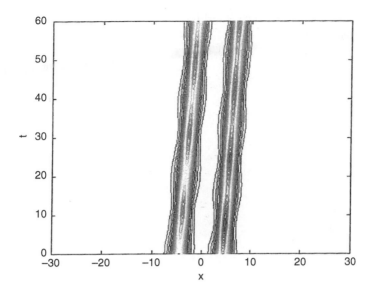

FIGURE 8.6
SSI with $v = 0.3$, $\alpha = 0.8$, $\gamma = -0.35$, $\sigma = 0.03$.

where

$$D_l \equiv \chi(A_l) = A_l^p \left(\frac{4p^2}{1+p} \right)^{\frac{1}{2p}} \tag{8.219}$$

Here (8.66) for dual-power law modifies to

$$i\frac{\partial q_l}{\partial t} + \frac{1}{2}\frac{\partial^2 q_l}{\partial x^2} = i\epsilon R[q_l, q_l^*]$$

$$- \left[\sum_{r=0}^{p} \binom{p}{r} q_1^{p-r} q_2^r \right] \left[\sum_{r=0}^{p} \binom{p}{r} (q_1^*)^{p-r} (q_2^*)^r \right] (q_1 + q_2)$$

$$- v \left[\sum_{r=0}^{2p} \binom{2p}{r} q_1^{2p-r} q_2^r \right] \left[\sum_{r=0}^{2p} \binom{2p}{r} (q_1^*)^{2p-r} (q_2^*)^r \right] (q_1 + q_2) \tag{8.220}$$

where the separation

$$|q|^{2p}q + v|q|^{4p}q = \left[\sum_{r=0}^{p} \binom{p}{r} q_1^{p-r} q_2^r \right] \left[\sum_{r=0}^{p} \binom{p}{r} (q_1^*)^{p-r} (q_2^*)^r \right] (q_1 + q_2)$$

$$+ v \left[\sum_{r=0}^{2p} \binom{2p}{r} q_1^{2p-r} q_2^r \right] \left[\sum_{r=0}^{2p} \binom{2p}{r} (q_1^*)^{2p-r} (q_2^*)^r \right] (q_1 + q_2) \tag{8.221}$$

was used based on the degree of overlapping. By virtue of the SPT, the evolution equations of the soliton parameters are

$$\frac{d A_l}{dt} = F_1^{(l)}(A, \Delta x, \Delta\phi; v, p) + \epsilon M_l \tag{8.222}$$

$$\frac{d B_l}{dt} = F_2^{(l)}(A, \Delta x, \Delta\phi; v, p) + \epsilon N_l \tag{8.223}$$

$$\frac{d T_l}{dt} = -B_l - F_3(A, \Delta x, \Delta\phi; v, p) + \epsilon Q_l \tag{8.224}$$

and

$$\frac{d\delta_l}{dt} = \frac{A_l^{2p}}{2p+2} + \frac{B_l^2}{2} + F_4(A, \Delta x, \Delta\phi; v, p) + \epsilon P_l \tag{8.225}$$

where for the case of dual-power law

$$M_l = h_1^{(2)}(A_l) \int_{-\infty}^{\infty} \Re\left\{\hat{R}[q_l, q_l^*]e^{-i\phi_l}\right\} \frac{d\tau_l}{(1 + a_l \cosh \tau_l)^{\frac{1}{2p}}} d\tau_l \tag{8.226}$$

$$N_l = h_2^{(2)}(A_l) \int_{-\infty}^{\infty} \Im\left\{\hat{R}[q_l, q_l^*]e^{-i\phi_l}\right\} \frac{\sinh \tau_l}{(1 + a_l \cosh \tau_l)^{\frac{1}{2p}}} d\tau_l \tag{8.227}$$

$$Q_l = h_3^{(2)}(A_l) \int_{-\infty}^{\infty} \Re\left\{\hat{R}[q_l, q_l^*]e^{-i\phi_l}\right\} \frac{\tau_l}{(1 + a_l \cosh \tau_l)^{\frac{1}{2p}}} d\tau_l \tag{8.228}$$

and

$$P_l = h_4^{(2)}(A_l) \int_{-\infty}^{\infty} \Im\left\{\hat{R}[q_l, q_l^*]e^{-i\phi_l}\right\} \frac{(1 - a_l \tau_l \sinh \tau_l)}{(1 + a_l \cosh \tau_l)^{\frac{1}{2p}}} d\tau_l \tag{8.229}$$

In addition, the following notations are used

$$\hat{R}[q_l, q_l^*] = R[q_l, q_l^*] + i|q_l|^{2p}q_l + i\nu|q_l|^{4p}q_l$$

$$-i\left[\sum_{r=0}^{p} \binom{p}{r} q_1^{p-r}q_2^r\right]\left[\sum_{r=0}^{p} \binom{p}{r} (q_1^*)^{p-r}(q_2^*)^r\right](q_1 + q_2)$$

$$-i\left[\sum_{r=0}^{2p} \binom{2p}{r} q_1^{2p-r}q_2^r\right]\left[\sum_{r=0}^{2p} \binom{2p}{r} (q_1^*)^{2p-r}(q_2^*)^r\right](q_1 + q_2)$$

$$\tag{8.230}$$

For the case of dual-power law nonlinearity, the study is split into two subsections.

8.3.4.1 Non-Hamiltonian Perturbations

In the presence of nonperturbation terms, as given by (8.2), the dynamical system of the soliton parameters by virtue of the SPT are

$$
\frac{dA}{dt} = \frac{4\epsilon\delta A^{2m+2}}{2^{\frac{m+1}{p}}a^{\frac{m+1}{p}}D} F\left(\frac{m+1}{p}, \frac{m+1}{p}, \frac{m+1}{p} + \frac{1}{2}; \frac{a-1}{2a}\right) B\left(\frac{m+1}{p}, \frac{1}{2}\right)
$$

$$
+ \frac{2\epsilon\sigma A^4}{D^2} \int_{-\infty}^{\infty} \frac{1}{(1+a\cosh\tau)^{\frac{1}{p}}} \left(\int_{-\infty}^{\tau} \frac{ds}{(1+a\cosh s)^{\frac{1}{p}}}\right) d\tau
$$

$$
- \frac{4\epsilon\beta A^2}{Da^{\frac{1}{p}}2^{\frac{1}{p}}} \left[D^2 F\left(2+\frac{1}{p}, \frac{1}{p}, \frac{3}{2} + \frac{1}{p}; \frac{a-1}{2a}\right) B\left(\frac{1}{p}, \frac{3}{2}\right) \right.
$$

$$
\left. + B^2 F\left(\frac{1}{p}, \frac{1}{p}, \frac{1}{2} + \frac{1}{p}; \frac{a-1}{2a}\right) B\left(\frac{1}{p}, \frac{1}{2}\right) \right] \tag{8.231}
$$

$$
\frac{dB}{dt} = -\frac{\epsilon\beta B D^2}{4p^2 A^2} \frac{F\left(2+\frac{1}{p}, 1+\frac{1}{p}, 2+\frac{1}{p}; \frac{a-1}{2a}\right)}{F\left(\frac{1}{p}, \frac{1}{p}, \frac{1}{2} + \frac{1}{p}; \frac{a-1}{2a}\right)} \frac{B\left(1+\frac{1}{p}, 1\right)}{B\left(\frac{1}{p}, \frac{1}{2}\right)} \tag{8.232}
$$

For the fixed point of the dynamical system, given by (8.172) and (8.173), with $A = 1$ and $B = 0$, one recovers

$$
\beta = \frac{\delta}{2^{\frac{m}{p}}a^{\frac{m}{p}}} \left(\frac{1+p}{2p^2}\right)^{\frac{1}{p}} \frac{F\left(\frac{m+1}{p}, \frac{m+1}{p}, \frac{m+1}{p} + \frac{1}{2}; \frac{a-1}{2a}\right)}{F\left(2+\frac{1}{p}, \frac{1}{p}, \frac{3}{2} + \frac{1}{p}; \frac{a-1}{2a}\right)} \frac{B\left(\frac{m+1}{p}, \frac{3}{2}\right)}{B\left(\frac{1}{p}, \frac{3}{2}\right)} + \frac{\sigma a^{\frac{1}{p}}}{2^{\frac{p-1}{p}}} \left(\frac{1+p}{2p^2}\right)^{\frac{3}{2p}}
$$

$$
\frac{1}{B\left(\frac{1}{p}, \frac{3}{2}\right) F\left(2+\frac{1}{p}, \frac{1}{p}, \frac{3}{2} + \frac{1}{p}; \frac{a-1}{2a}\right)} \int_{-\infty}^{\infty} \frac{1}{(1+a\cosh\tau)^{\frac{1}{p}}}
$$

$$
\left(\int_{-\infty}^{\tau} \frac{ds}{(1+a\cosh s)^{\frac{1}{p}}}\right) d\tau \tag{8.233}
$$

Thus, using (8.79)–(8.80), one can obtain

$$
\frac{d^2(\Delta x)}{dt^2} + \epsilon\beta G \frac{d(\Delta x)}{dt} + F_2^{(1)} - F_2^{(2)} = 0 \tag{8.234}
$$

where β is given by (8.233) and $G > 0$ represents the coefficient of $-\epsilon\beta\Delta B$ in $d(\Delta B)/dt = dB_1/dt - dB_2/dt$. Now, (8.234) shows damping in the separation of solitons, thus proving that there will be a suppression of SSI in the presence of the perturbation terms given by (8.2). Thus, in Figure 8.7, the numerical simulations show that the suppression of SSI is achieved for dual-power law as proved in the QPT.

8.3.4.2 Hamiltonian Perturbations

In the presence of the perturbation terms given by (8.4), the dynamical system of the soliton parameters by virtue of the SPT are

$$
\frac{dA}{dt} = 0 \tag{8.235}
$$

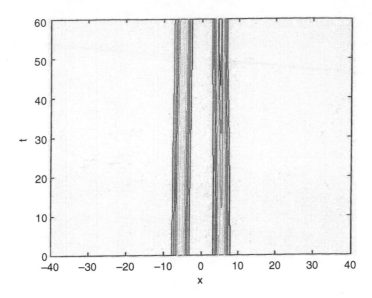

FIGURE 8.7
SSI with $m = 1$, $\delta = \sigma = 0.001$.

$$\frac{dB}{dt} = 0 \qquad (8.236)$$

and

$$\frac{dx_0}{dt} = -B - \epsilon(\mu + 3\gamma B^2)$$

$$-3\frac{\epsilon\gamma D^2}{2p^2}\frac{F\left(2+\frac{1}{p},\frac{1}{p},\frac{1}{p}+\frac{3}{2};\frac{a-1}{2a}\right)}{F\left(\frac{1}{p},\frac{1}{p},\frac{1}{2}+\frac{1}{p};\frac{a-1}{2a}\right)}\frac{B\left(\frac{1}{p},\frac{1}{2}\right)}{B\left(\frac{1}{p},\frac{3}{2}\right)}$$

$$-\epsilon(3\lambda + 2\mu)\frac{A^2}{2^{\frac{1}{p}}a^{\frac{1}{p}}}\frac{F\left(\frac{2}{p},\frac{2}{p},\frac{2}{p}+\frac{1}{2};\frac{a-1}{2a}\right)}{F\left(\frac{1}{p},\frac{1}{p},\frac{1}{2}+\frac{1}{p};\frac{a-1}{2a}\right)}\frac{B\left(\frac{1}{p},\frac{1}{2}\right)}{B\left(\frac{2}{p},\frac{1}{2}\right)} \qquad (8.237)$$

From (8.79), (8.80), (8.83), and (8.84), one can now conclude that

$$\frac{d(\Delta A)}{dt} = F_1^{(1)}(A, \Delta x, \Delta\phi) - F_1^{(2)}(A, \Delta x, \Delta\phi) \qquad (8.238)$$

$$\frac{d(\Delta B)}{dt} = F_2^{(1)}(A, \Delta T, \Delta\phi) - F_2^{(2)}(A, \Delta x, \Delta\phi) \qquad (8.239)$$

$$\frac{d(\Delta x)}{dt} = \frac{dx_1}{dt} - \frac{dx_2}{dt} \qquad (8.240)$$

and

$$\frac{d(\Delta\phi)}{dt} = \frac{1}{2}A\Delta A \qquad (8.241)$$

For in-phase injection of solitons with unequal amplitudes

$$A = \frac{1}{2}(A_0 + 1) \tag{8.242}$$

$$B = 0 \tag{8.243}$$

$$\Delta A_0 = A_0 - 1 \tag{8.244}$$

$$\Delta B_0 = 0 \tag{8.245}$$

$$\Delta T_0 = T_0 \tag{8.246}$$

$$\Delta \phi_0 = 0 \tag{8.247}$$

and

$$\Delta \phi = \Delta \delta \tag{8.248}$$

so that from (8.237), one can get

$$\frac{d(\Delta x)}{dt} = -\Delta B + \epsilon G\left(\alpha, \lambda, \mu, \gamma, \sigma\right) g \left\{ \frac{d(\Delta \phi)}{dt} \right\} \tag{8.249}$$

where G is the functional form that depends on the said parameters. For in-phase injection of solitons with unequal amplitudes, where $\Delta B = 0$

$$\Delta x = x_0 + \epsilon G\left(\alpha, \lambda, \mu, \gamma, \sigma\right) h \left\{ \frac{d(\Delta \phi)}{dt} \right\} \tag{8.250}$$

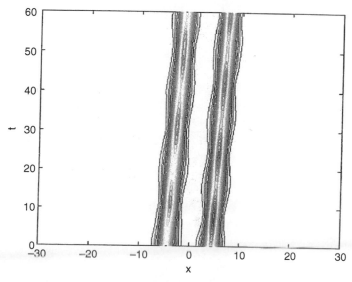

FIGURE 8.8
SSI with $v = 0.5$, $p = 1/2$, $\alpha = 0.8$, $\gamma = -0.25$, $\sigma = 0.05$, $\mu = 0.5$.

where

$$h_j(s) = \int g_j(s)ds \tag{8.251}$$

for $j = 1, 2$. Thus

$$\Delta x = x_0 + O(\epsilon) \tag{8.252}$$

Now, $x_0 \sim O(1)$ so that $\Delta x \nrightarrow 0$ and, thus, the pulses do not collide during the transmission, as observed in the numerical simulations in Figure 8.8.

Exercises

1. Use the adiabatic parameter dynamics given by (8.116), (8.117), (8.109), and (8.110) to obtain the relations given by (8.118)–(8.121).
2. Use the adiabatic parameter dynamics given by (8.155), (8.156), (8.148), and (8.149) to obtain the relations given by (8.157)–(8.160).
3. For the fixed point of the dynamical system given by (8.195) and (8.196), prove that β is given by (8.198) for $A = 1$ and $B = 0$.
4. For the fixed point of the dynamical system given by (8.231) and (8.232), prove that β is given by (8.233) for $A = 1$ and $B = 0$.

9

Stochastic Perturbation

The use of optical amplifiers affects the evolution of solitons considerably. The reason is that, although amplifiers are necessary to restore the soliton energy, they add noise originating from amplified spontaneous emission (ASE). The effect of ASE is to change the soliton parameters randomly. Variances of such fluctuations can be calculated by treating ASE as a perturbation. In this chapter, such calculations will be carried out for Kerr law, power law, parabolic law, and dual-power law fibers.

9.1 Introduction

Besides deterministic-type perturbations, one also needs to take into account, for practical considerations, stochastic-type perturbations. These effects can be classified into three basic types:

1. Stochasticity associated with the chaotic nature of the initial pulse due to partial coherence of the laser-generated radiation.

2. Stochasticity due to random nonuniformities in the optical fibers, such as fluctuations in the values of the dielectric constant, random variations of the fiber diameter, and more.

3. The chaotic field caused by dynamic stochasticity might arise from a periodic modulation of the system parameters or when a periodic array of pulses propagates in a fiber-optic resonator.

Thus, stochasticity is inevitable in optical soliton communications.

Stochasticity is basically of two types, namely homogenous and nonhomogenous [89]:

- In the homogenous case, stochasticity originates due to random perturbation of the fiber, such as the density fluctuation of the fiber material or random variations in the fiber diameter.

- In the inhomogenous case, stochasticity is present in the input pulse of the fiber. So the parameter dynamics are deterministic, although the initial values are random.

Considering the effects of perturbation [89] on the propagation of solitons through optical fibers for the $1+1$ dimensional case, the perturbed nonlinear Schrödinger's equation is

$$iq_t + \frac{1}{2}q_{xx} + F(|q|^2)q = i \epsilon R \tag{9.1}$$

with

$$R = \delta|q|^{2m}q + \beta q_{xx} + \sigma(x, t) \tag{9.2}$$

where $\sigma(x, t)$ is the random stochastic perturbation term.

The amplifiers that are placed along soliton transmission lines to restore soliton energy introduce noise that originates from ASE. To study the impact of noise on soliton evolution, the evolution of the mean free velocity of the soliton due to ASE will be studied in this chapter. In the case of lumped amplification, solitons are perturbed by ASE in a discrete fashion at the location of the amplifiers. It can be assumed that noise is distributed all along the fiber length since the amplifier spacing satisfies $z_a \ll 1$. In (9.2), $\sigma(x, t)$ represents the Markovian stochastic process with Gaussian statistics and assumes that $\sigma(x, t)$ [227] is a function of t only so that $\sigma(x, t) = \sigma(t)$. Now, the complex stochastic term

$$\sigma(t) = \sigma_1(t) + i\sigma_2(t) \tag{9.3}$$

is further assumed to be independently delta correlated in both $\sigma_1(t)$ and $\sigma_2(t)$ with

$$\langle \sigma_1(t) \rangle = \langle \sigma_2(t) \rangle = \langle \sigma_1(t)\sigma_2(t') \rangle = 0 \tag{9.4}$$

$$\langle \sigma_1(t)\sigma_1(t') \rangle = 2D_1\delta(t - t') \tag{9.5}$$

$$\langle \sigma_2(t)\sigma_2(t') \rangle = 2D_2\delta(t - t') \tag{9.6}$$

Now, in (9.5) and (9.6), if it is assumed that $D_1 = D_2 = D$, then

$$\langle \sigma(t) \rangle = 0 \tag{9.7}$$

and

$$\langle \sigma(t)\sigma(t') \rangle = 2D\delta(t - t') \tag{9.8}$$

In soliton units, one gets

$$D = \frac{F_n F_G}{N_{ph} z_a} \tag{9.9}$$

where F_n is the amplifier noise figure, while

$$F_G = \frac{(G - 1)^2}{G \ln G} \tag{9.10}$$

is related to the amplifier gain G and, finally, N_{ph} is the average number of photons in the pulse propagating as a fundamental soliton.

In the presence of these perturbation terms, the adiabatic dynamics of the conserved quantities and the soliton parameter κ can be computed from (1.34), (1.35), and (1.37) using the R that is given by (9.2). Also, the general form of the soliton given by (2.24) is used so that

$$\frac{dE}{dt} = \frac{2\epsilon}{\beta}[\delta A^{2m+2} I_{0,2m+2,0,0} + \beta A^2 B(B^2 I_{0,1,0,1} - \kappa^2 I_{0,2,0,0})]$$

$$+ 2\epsilon A \int_{-\infty}^{\infty} g(\tau)(\sigma_1 \cos\phi + \sigma_2 \sin\phi) \, dx \qquad (9.11)$$

and

$$\frac{dM}{dt} = \frac{2\epsilon}{B}[\kappa A^2(2\beta AB^3 I_{0,0,3,0} - \delta A^{2m} I_{0,2m+2,0,0}) + \beta\kappa A^2(\kappa^2 I_{0,2,0,0} - B^2 I_{0,1,0,1})]$$

$$+ 2\epsilon A \int_{-\infty}^{\infty} \left[g(\tau)(\sigma_1 \cos\phi + \sigma_2 \sin\phi) - B\frac{dg}{d\tau}(\sigma_2 \cos\phi - \sigma_1 \sin\phi) \right] dx$$

$$(9.12)$$

From (9.11) and (9.12), it is possible to derive

$$\frac{d\kappa}{dt} = \frac{4\epsilon\beta\kappa B}{I_{0,2,0,0}}(\kappa^2 I_{0,2,0,0} + B^2 I_{0,0,2,0} - B^2 I_{0,1,0,1})$$

$$- \frac{2\epsilon B}{A I_{0,2,0,0}} \int_{-\infty}^{\infty} \left[B\frac{dg}{d\tau}(\sigma_2 \cos\phi - \sigma_1 \sin\phi) + 2\kappa g(\tau)(\sigma_1 \cos\phi + \sigma_2 \sin\phi) \right] dx$$

$$(9.13)$$

The study of solitons in the presence of stochastic perturbation will now be split into the following four sections: Kerr law nonlinearity, power law, parabolic law, and dual-power law.

9.2   Kerr Law

For the Kerr law nonlinearity equations (9.11) and (9.13), the perturbation terms given by (9.2) respectively modify to

$$\frac{dA}{dt} = \epsilon \left[A \left\{ \delta A^{2m} \frac{\Gamma\left(\frac{1}{2}\right)\Gamma(m+1)}{\Gamma\left(m+\frac{3}{2}\right)} - \frac{2}{3}\beta A^2 - 3\beta\kappa^2 \right\} \right.$$

$$\left. + \pi \left\{ \frac{\sigma_1 \cos(\omega t + \sigma_0) - \sigma_2 \sin(\omega t + \sigma_0)}{\cosh\left(\frac{\pi\kappa}{2\eta}\right)} \right\} \right] \qquad (9.14)$$

and

$$\frac{d\kappa}{dt} = -\epsilon \left[\frac{4}{3}\beta A^2 \kappa + \frac{\pi}{2\eta}(2A^2 + 2\kappa^2 - \kappa A) \right.$$
$$\left. \left\{ \frac{\sigma_1 \cos(\omega t + \sigma_0) - \sigma_2 \sin(\omega t + \sigma_0)}{\cosh\left(\frac{\pi\kappa}{2A}\right)} \right\} \right] \tag{9.15}$$

Equations (9.14) and (9.15) are difficult to analyze. If the terms with σ_1 and σ_2 are suppressed, the resulting dynamical system has a stable fixed point, namely a sink given by $(\bar{A}, \bar{\kappa}) = (\tilde{A}, 0)$ where

$$\bar{A} = \left[\frac{2\beta\Gamma\left(m + \frac{3}{2}\right)}{3\delta\Gamma\left(\frac{1}{2}\right)\Gamma(m+1)} \right]^{\frac{1}{2m-2}} \tag{9.16}$$

Now, linearizing the dynamical system about this fixed point and simplifying gives

$$\frac{dA}{dt} = -\epsilon\left(A^{2m+1} - \frac{\zeta}{\bar{A}} \right) \tag{9.17}$$

and

$$\frac{d\kappa}{dt} = -\epsilon[\kappa - \zeta\,(1 + A - \kappa)] \tag{9.18}$$

where

$$\zeta = \pi \left\{ \frac{\sigma_1 \cos\,(\omega t + \sigma_0) - \sigma_2 \sin\,(\omega t + \sigma_0)}{\cosh\left(\frac{\pi\kappa}{2A}\right)} \right\} \tag{9.19}$$

Equations (9.17) and (9.18) are called the *Langevin equations* and will now be analyzed to compute the soliton mean drift velocity. If soliton parameters are chosen such that ζA is small, then (9.18) yields

$$\frac{d\kappa}{dt} = -\epsilon[\kappa - \zeta(1 - \kappa)] \tag{9.20}$$

One can solve (9.20) for κ, and eventually the mean drift velocity of the soliton can be obtained. The stochastic phase factor of the soliton is defined by

$$\psi(t, y) = \int_y^t \zeta(s)ds \tag{9.21}$$

where $t > y$. Assuming that σ is a Gaussian stochastic variable, one arrives at

$$\langle e^{\psi(t,y)} \rangle = e^{D(t-y)} \tag{9.22}$$

$$\langle e^{[\psi(t,y)+\psi(t',y')]} \rangle = e^{D\theta} \tag{9.23}$$

where

$$\theta = 2(t + t' - y - y') - |t - t'| - |y - y'| \tag{9.24}$$

$$\langle \zeta(y)e^{-\psi(t,y)} \rangle = \frac{\partial}{\partial y}\langle e^{-\psi(t,y)} \rangle = De^{D(t-y)} \tag{9.25}$$

and

$$\langle \zeta(y)\zeta(y')e^{[-\psi(t,y)-\psi(t',y')]} \rangle = 2D\delta(y - y')e^{D\theta} + \frac{\partial^2}{\partial y \partial y'}e^{D\theta} \tag{9.26}$$

Now, solving (9.20) with the initial condition as $\kappa(0) = 0$ and using equations (9.21)–(9.26), the soliton mean drift velocity is given by

$$\langle \kappa(t) \rangle = -\frac{D}{1-D}\left\{1 - e^{-\epsilon(1-D)t}\right\} \tag{9.27}$$

From (9.27), it follows that

$$\lim_{t \to \infty} \langle \kappa(t) \rangle = -\frac{D}{1-D} \tag{9.28}$$

so that, in the limiting case, the mean free velocity of the soliton is given as

$$\langle v \rangle = \frac{D}{1-D} \tag{9.29}$$

Thus, for large t, $\langle v(t) \rangle$ approaches a constant value provided $D < 1$. For $D > 1$, $\langle \kappa(t) \rangle$ becomes unbounded for large t.

9.3 Power Law

For the case of power law nonlinearity, equations (9.11) and (9.13), respectively, modify to

$$\begin{aligned}
\frac{dA}{dt} &= \frac{2\epsilon\delta}{2-p}A^{2m+1}\left(\frac{1+p}{2p^2}\right)^{\frac{1}{2p}}\frac{\Gamma\left(\frac{1}{p}+\frac{1}{2}\right)}{\Gamma\left(\frac{1}{p}\right)}\frac{\Gamma\left(\frac{m+1}{p}\right)}{\Gamma\left(\frac{m+1}{p}+\frac{1}{2}\right)} \\
&\quad + \frac{2\epsilon\beta}{2-p}\frac{A^{p-1}}{B}\left(\frac{2p^2}{p+1}\right)^{\frac{p-1}{2p}}\frac{\Gamma\left(\frac{1}{p}+\frac{1}{2}\right)}{\Gamma\left(\frac{1}{p}\right)} \\
&\quad \left[\frac{A^2 B^2}{p^2}\frac{\Gamma\left(\frac{p+1}{p}\right)}{\Gamma\left(\frac{p+1}{p}+\frac{1}{2}\right)} - \frac{A^2}{p^2}(\kappa^2 p^2 + B^2)\frac{\Gamma\left(\frac{1}{p}\right)}{\Gamma\left(\frac{1}{p}+\frac{1}{2}\right)}\right] \\
&\quad + \frac{2\epsilon A}{2-p}A^{p-1}\left(\frac{2p^2}{1+p}\right)^{\frac{p-1}{2p}}\frac{\Gamma\left(\frac{1}{p}+\frac{1}{2}\right)}{\Gamma\left(\frac{1}{2}\right)\Gamma\left(\frac{1}{p}\right)} \\
&\quad \left[\sigma_1\int_{-\infty}^{\infty}\frac{\cos\phi}{\cosh^{\frac{1}{p}}\tau}dx + \sigma_2\int_{-\infty}^{\infty}\frac{\sin\phi}{\cosh^{\frac{1}{p}}\tau}dx\right]
\end{aligned} \tag{9.30}$$

and

$$\frac{d\kappa}{dt} = \frac{4\,\epsilon\,\beta}{p^2}\kappa\,A^2 B^{\frac{2p-2}{p}}\left(\frac{2p^2}{p+1}\right)^{\frac{1}{2}}\left(\frac{p-2}{p+2}\right)$$

$$+ 4\,\epsilon\,\kappa\,AB^{\frac{p-2}{p}}\left(\frac{2p^2}{1+p}\right)^{\frac{1}{p}}\frac{\Gamma\left(\frac{1}{p}+\frac{1}{2}\right)}{\Gamma\left(\frac{1}{2}\right)\Gamma\left(\frac{1}{p}\right)}\int_{-\infty}^{\infty}\left[\frac{B}{p}\frac{\tanh\tau}{\cosh^{\frac{1}{p}}\tau}(\sigma_2\cos\phi-\sigma_1\sin\phi)\right.$$

$$\left.+\frac{2\kappa}{\cosh^{\frac{1}{p}}\tau}(\sigma_1\cos\phi+\sigma_2\sin\phi)\right]dx \tag{9.31}$$

Equations (9.30) and (9.31) are difficult to analyze. If the terms with σ_1 and σ_2 are suppressed, the resulting dynamical system has a stable fixed point, namely a sink given by $(\bar{A}, \bar{\kappa}) = (\bar{A}, 0)$ where

$$\bar{A} = \left[\frac{2\beta}{\delta(p+1)}\frac{\Gamma\left(\frac{m+1}{p}+\frac{1}{2}\right)}{\Gamma\left(\frac{m+1}{p}\right)}\left\{\frac{\Gamma\left(\frac{p+1}{p}\right)}{\Gamma\left(\frac{p+1}{p}+\frac{1}{2}\right)}-\frac{\Gamma\left(\frac{1}{p}\right)}{\Gamma\left(\frac{1}{p}+\frac{1}{2}\right)}\right\}\right]^{\frac{1}{2(m-p)}} \tag{9.32}$$

After simplification, linearizing the dynamical system about this fixed point gives

$$\frac{dA}{dt} = -\epsilon\left(A^{2m+1}-\frac{\zeta_1^{(1)}}{\bar{A}}\right) \tag{9.33}$$

and

$$\frac{d\kappa}{dt} = -\epsilon\left[\kappa-\zeta_2^{(1)}(1+A-\kappa)\right] \tag{9.34}$$

where

$$\zeta_1^{(1)} = \sigma_1\int_{-\infty}^{\infty}\frac{\cos\phi}{\cosh^{\frac{1}{p}}\tau}dx+\sigma_2\int_{-\infty}^{\infty}\frac{\sin\phi}{\cosh^{\frac{1}{p}}\tau}dx \tag{9.35}$$

and

$$\zeta_2^{(1)} = \int_{-\infty}^{\infty}\left[\frac{B}{p}\frac{\tanh\tau}{\cosh^{\frac{1}{p}}\tau}(\sigma_2\cos\phi-\sigma_1\sin\phi)+\frac{2\kappa}{\cosh^{\frac{1}{p}}\tau}(\sigma_1\cos\phi+\sigma_2\sin\phi)\right]dx \tag{9.36}$$

Similarly, as in the case of Kerr law nonlinearity, these Langevin equations lead to the mean drift velocity of the soliton as

$$\langle v\rangle = \frac{D}{1-D} \tag{9.37}$$

9.4 Parabolic Law

For parabolic law nonlinearity, (9.11) and (9.13) respectively modify to

$$
\frac{dA}{dt} = \frac{\epsilon \delta A^{2m+1}}{2^m a^{m+1}} F\left(m+1, m+1, m+\frac{3}{2}; \frac{a-1}{2a}\right) B\left(m+1, \frac{1}{2}\right)
$$
$$
+ \frac{\epsilon \sqrt{2}\beta A}{a}\left[\kappa^2 F\left(1, 1, \frac{3}{2}; \frac{a-1}{2a}\right) B\left(1, \frac{1}{2}\right)\right.
$$
$$
\left. - 2A^2 F\left(3, 1, \frac{5}{2}; \frac{a-1}{2a}\right) B\left(1, \frac{3}{2}\right)\right]
$$
$$
+ \epsilon a^2 A\sqrt{2}\left[\sigma_1 \int_{-\infty}^{\infty} \frac{\cos\phi}{(1+a\cosh\tau)^{\frac{1}{2}}} dx \right.
$$
$$
\left. + \sigma_2 \int_{-\infty}^{\infty} \frac{\sin\phi}{(1+a\cosh\tau)^{\frac{1}{2}}} dx\right] \tag{9.38}
$$

and

$$
\frac{d\kappa}{dt} = -\frac{\epsilon\beta\kappa A^2}{2} \frac{F\left(3, 2, 3; \frac{a-1}{2a}\right)}{F\left(1, 1, \frac{3}{2}; \frac{a-1}{2a}\right)} \frac{B(2, 1)}{B\left(1, \frac{1}{2}\right)}
$$
$$
- \frac{\epsilon\sqrt{2}}{AE} \int_{-\infty}^{\infty}\left[\frac{a B (\sigma_2 \cos\phi - \sigma_1 \sin\phi)\sinh\tau}{(1+a\cosh\tau)^{\frac{3}{2}}} - \frac{2\kappa (\sigma_1 \cos\phi + \sigma_2 \sin\phi)}{(1+a\cosh\tau)^{\frac{1}{2}}}\right] dx \tag{9.39}
$$

After simplification, linearizing the dynamical system about this fixed point gives

$$
\frac{dA}{dt} = -\epsilon\left(A^{2m+1} - \frac{\zeta_1^{(2)}}{\bar{A}}\right) \tag{9.40}
$$

and

$$
\frac{d\kappa}{dt} = -\epsilon\left[\kappa - \zeta_2^{(2)}(1 + A - \kappa)\right] \tag{9.41}
$$

where $\bar{A}$ is the fixed point of the amplitude, while

$$
\zeta_1^{(2)} = \sigma_1 \int_{-\infty}^{\infty} \frac{\cos\phi}{(1+a\cosh\tau)^{\frac{1}{2}}} dx + \sigma_2 \int_{-\infty}^{\infty} \frac{\sin\phi}{(1+a\cosh\tau)^{\frac{1}{2}}} dx \tag{9.42}
$$

and

$$
\zeta_2^{(2)} = \int_{-\infty}^{\infty}\left[\frac{a B(\sigma_2 \cos\phi - \sigma_1 \sin\phi)\sinh\tau}{(1+a\cosh\tau)^{\frac{3}{2}}} - \frac{2\kappa(\sigma_1 \cos\phi + \sigma_2 \sin\phi)}{(1+a\cosh\tau)^{\frac{1}{2}}}\right] dx \tag{9.43}
$$

Again, as in the case of Kerr law nonlinearity, these Langevin equations lead to the mean drift velocity of the parabolic law soliton as

$$\langle v \rangle = \frac{D}{1 - D} \tag{9.44}$$

9.5 Dual-Power Law

In the case of dual-power law nonlinearity, (9.11) and (9.13), respectively, modify to

$$
\begin{aligned}
\frac{dA}{dt} &= \frac{4\epsilon\delta A^{2m+2}}{Bb^{\frac{m+1}{p}} 2^{\frac{m+1}{p}}} F\left(\frac{m+1}{p}, \frac{m+1}{p}, \frac{m+1}{p} + \frac{1}{2}; \frac{b-1}{2b}\right) B\left(\frac{m+1}{p}, \frac{1}{2}\right) \\
&\quad - \frac{4\epsilon\beta A^2}{Bb^{\frac{1}{p}} 2^{\frac{1}{p}}} \left[B^2 F\left(2 + \frac{1}{p}, \frac{1}{p}, \frac{3}{2} + \frac{1}{p}; \frac{b-1}{2b}\right) B\left(\frac{1}{p}, \frac{3}{2}\right) \right. \\
&\quad \left. + \kappa^2 F\left(\frac{1}{p}, \frac{1}{p}, \frac{1}{2} + \frac{1}{p}; \frac{b-1}{2b}\right) B\left(\frac{1}{p}, \frac{1}{2}\right) \right] \\
&\quad + \frac{2\epsilon}{pLA^{p-2}} \left(\frac{p+1}{2p^2}\right)^{\frac{1}{2p}} \left[\sigma_1 \int_{-\infty}^{\infty} \frac{\cos\phi}{(1 + a\cosh\tau)^{\frac{1}{2p}}} dx \right. \\
&\quad \left. + \sigma_2 \int_{-\infty}^{\infty} \frac{\sin\phi}{(1 + a\cosh\tau)^{\frac{1}{2p}}} dx \right] \tag{9.45}
\end{aligned}
$$

and

$$
\begin{aligned}
\frac{d\kappa}{dt} &= -\frac{\epsilon\beta\kappa B^2}{4p^2 A^2} \frac{F\left(2 + \frac{1}{p}, 1 + \frac{1}{p}, 2 + \frac{1}{p}; \frac{b-1}{2b}\right)}{F\left(\frac{1}{p}, \frac{1}{p}, \frac{1}{2} + \frac{1}{p}; \frac{b-1}{2b}\right)} \frac{B\left(1 + \frac{1}{p}, 1\right)}{B\left(\frac{1}{p}, \frac{1}{2}\right)} \\
&\quad - \frac{\epsilon}{E} \int_{-\infty}^{\infty} \left[\frac{aB(\sigma_2\cos\phi - \sigma_1\sin\phi)\sinh\tau}{2p (1 + a\cosh\tau)^{\frac{2p+1}{2p}}} - \frac{2\kappa(\sigma_1\cos\phi + \sigma_2\sin\phi)}{(1 + a\cosh\tau)^{\frac{1}{2p}}} \right] dx \tag{9.46}
\end{aligned}
$$

After simplification, linearizing the dynamical system about this fixed point gives

$$\frac{dA}{dt} = -\epsilon\left(A^{2m+1} - \frac{\varsigma_1^{(3)}}{\bar{A}}\right) \tag{9.47}$$

and

$$\frac{d\kappa}{dt} = -\epsilon\left[\kappa - \varsigma_2^{(3)}(1 + A - \kappa)\right] \tag{9.48}$$

where $\bar{A}$ is the fixed point of the amplitude, while

$$\zeta_1^{(3)} = \sigma_1 \int_{-\infty}^{\infty} \frac{\cos\psi}{(1+a\cosh\tau)^{\frac{1}{2p}}} dx + \sigma_2 \int_{-\infty}^{\infty} \frac{\sin\phi}{(1+a\cosh\tau)^{\frac{1}{2p}}} dx \quad (9.49)$$

and

$$\zeta_2^{(3)} = \int_{-\infty}^{\infty} \left[\frac{a B (\sigma_2 \cos\phi - \sigma_1 \sin\phi)\sinh\tau}{2p \quad (1+a\cosh\tau)^{\frac{2p+1}{2p}}} - \frac{2\kappa (\sigma_1 \cos\phi + \sigma_2 \sin\phi)}{(1+a\cosh\tau)^{\frac{1}{2p}}} \right] dx \quad (9.50)$$

Once again, as in the case of Kerr law nonlinearity, one can derive the mean drift velocity of the soliton as

$$\langle v \rangle = \frac{D}{1-D} \quad (9.51)$$

Exercises

1. Obtain the adiabatic parameter dynamics for the soliton amplitude and frequency that are given by (9.14) and (9.15).

2. For Kerr law nonlinearity, obtain the two-point correlation for

$$\langle \kappa(t)\kappa(t') \rangle$$

3. For power law nonlinearity, establish the fixed point of the amplitude from (9.30) and (9.31) for σ_1 and σ_2 suppressed.

10

Optical Couplers

This chapter provides an overview of optical couplers. In Section 10.1, an introduction to optical couplers is given. Also discussed in this section are the various kinds of couplers and their functions. In Section 10.2, the concept of a twin-core coupler is discussed. The various parameters of the solitons are defined and the corresponding parameter dynamics of the solitons are obtained from the corresponding nonlinear Schrödinger's equation (NLSE). In Section 10.3, multiple-core couplers are discussed. Finally, Section 10.4 contains a discussion on solitons in magneto-optic waveguides.

10.1 Introduction

In this chapter, the study of the NLSE will be extended to the case of optical couplers. Optical nonlinear couplers are very useful devices that distribute light from a main fiber into one or more branch fibers. Couplers also have applications as intensity-dependent switches and as limiters. They can be used to multiplex two incoming bit streams onto a fiber and also to demultiplex a single-bit stream. Optical couplers can be made as planar devices using semiconductor material or as dual-core, single-mode fibers with solitons propagating in each core. The coupling of energy from one guide to the other can occur due to the overlap of evanescent fields between the cores.

There are core, interaction-type couplers as well as surface-interaction-type couplers. In core-interaction-type couplers, the light energy transfer takes place through the core cross-section concatenating fibers or by using some form of imaging optics between the fibers (i.e., using lensing schemes such as rounded end fibers, a spherical lens to image the core of one fiber onto the core area of the other fiber, and a taper-ended fiber). In surface-interaction-type couplers, the light energy transfer takes place through the fiber surface and normal to the axis of the fiber by converting the guided core modes to cladding and refracted modes.

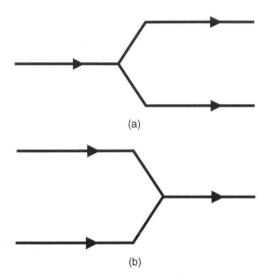

FIGURE 10.1
(a) Three port coupler as splitter, (b) three port coupler as combiner.

10.1.1 Types of Couplers and Their Functions

10.1.1.1 Three- and Four-Port Couplers

Figures 10.1(a) and 10.1(b) show the uses of a three-port coupler as splitter and combiner of signals. Light from the input fiber is coupled to the output fibers as shown in Figure 10.1(a), or the light from the branch fibers are combined to form a single input to the output fiber. In one method of splitting, called the *lateral offset method*, a single input fiber core is situated between the cores of two output fibers. In this method, the input power can be distributed in a well-defined proportion by appropriate control of the amount of lateral offset between the fibers.

Figure 10.2 shows the directional coupler, which is a four-port coupler. In this coupler, the fibers are generally twisted together and then spot fused under tension such that the fused section is elongated to form a biconical tapered structure. It can act as a three-port coupler (T coupler) if one of the input ends or one of the output ends is closed. As shown in the figure, each port is meant for a different function.

This type of coupler is based on the transfer of energy by surface interaction between the fibers. The amount of power taken from the main fiber or given to the main fiber depends on the length of the fused section of the fiber and the distance between the cores of the fused fibers. This can also act as a fiber laser amplifier.

10.1.1.2 Star Coupler or Multiport Couplers

A star coupler is used to distribute an optical signal from a single input fiber to multiple output fibers. Here, many fibers are bundled, twisted, heated, and pulled at the twisted area to get fiber fused biconical tapered star coupler.

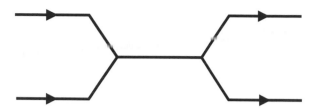

FIGURE 10.2
Four-port coupler.

10.1.2 Optical Switching

Solitons in nonlinear couplers have attracted much attention in recent years. The special interest is in optical switching because of the possibility of very short switching times in the femtosecond range. Switching is the process of energy redistribution between the cores for a given input. The problem of switching, although involved, can be accomplished when the stability of soliton states is known. Another practical application is in logic device or in high-bit rate communications in local networks. However, an optimal design of the fiber arrays is still awaited. Since solitons are used in long-distance fiber-optic communications it would be ideal if intensity-dependent switching could be performed in all optical devices, instead of converting from optical to electrical signals for electronic switching and then regenerating the optical signals. Moreover, fast optical computers can be designed using soliton switching and logic components involving couplers.

The switching of solitons can be realized in terms of all optical logic gates. Logic gate operations such as OR, AND, EX-OR; and NAND have been realized for soliton switching. It has been realized that these components will find useful applications in the field of optical information processing, permitting the realization of logic operations at speeds unattainable by conventional electronic systems. The ultimate aim should be the realization of all-optical digital computers interfaced, without the need of optoelectronic converters, of soliton communication systems in which the data rates are so high that conventional logic gates are too slow.

When a soliton pulse is transmitted through a birefringent fiber, an exchange of energy is possible between the pulses propagating in two orthogonal modes. The two orthogonally polarized solitons can trap one another and move at a common group velocity, in spite of their different modal indices (polarization dispersion). This phenomenon is known as *soliton trapping* and is quite important for optical soliton switching. This type of switching in fiber is intensity dependent. Depending upon its intensity, the pulse itself induces switching. The control of data flow in a fiber-optic communication system has been accomplished by switching of optical signals between fibers. Switching depends on the power and phase of the pulse.

In practice, an optical pulse is used for nonlinear switching. However, the switched pulse is severely distorted because only certain parts of the pulse have the right power for switching. In particular, pulse wings are too weak for switching to occur. Solitons can avoid pulse distortion because of their extraordinary stable property that the optical phase remains uniform across the entire pulse in spite of fiber nonlinearity. It has been predicted that the soliton interaction itself acts as a switch to suppress or enhance the switching dynamics.

10.2   Twin-Core Couplers

For twin-core couplers, wave propagation at relatively high field intensities is described by coupled nonlinear equations. In the dimensionless form, they are

$$iq_t + \frac{1}{2}q_{xx} + F(|q|^2)q = Kr \tag{10.1}$$

$$ir_t + \frac{1}{2}r_{xx} + F(|r|^2)r = Kq \tag{10.2}$$

The constant K represents the coupling coefficient between the cores of the fiber. In (10.1) and (10.2), q and r represent dimensionless forms of the optical fields in the respective cores of the optical fibers. This system of equations models various applications, such as intensity-dependent switches and devices for separating a compressed soliton from its broad "pedestal."

Equations (10.1) and (10.2) are not integrable due to the presence of the arbitrary functional form given by F and due to the coupling term. Therefore, the propagation of solitons through these twin-core optical fibers will be studied approximately. Also, equations (10.1) and (10.2), as it appears, do not have infinitely many integrals of motion. In fact, they have as few as three. They are respectively given by energy (E), linear momentum (M), and the Hamiltonian (H) that are respectively given by [31, 79]

$$E = \int_{-\infty}^{\infty} (|q|^2 + |r|^2)dx \tag{10.3}$$

$$M = \frac{i}{2} \int_{-\infty}^{\infty} \{(qq_x^* - q^*q_x) + (rr_x^* - r^*r_x)\}dx \tag{10.4}$$

and

$$H = \int_{-\infty}^{\infty} \left[\frac{1}{2}(|q_x|^2 + |r_x|^2) - f(|q|^2) - f(|r|^2) - K(qr^* + rq^*) \right] dx \tag{10.5}$$

where

$$f(I) = \int_0^I F(\xi)d\xi$$

and the intensity I is given by $I = |q|^2$ or $I = |r|^2$ depending on the core. For $K = 0$, the pulses in the two fibers are assumed to be of the following functional form

$$q(x, t) = A_1(t)g\left[B_1(t)\left(x - x_1(t)\right)\right]\exp\left[-i\kappa_1(t)(x - x_1(t)) + i\theta_1(t)\right] \qquad (10.6)$$

and

$$r(x, t) = A_2(t)g\left[B_2(t)\left(x - x_2(t)\right)\right]\exp\left[-i\kappa_1(t)(x - x_2(t)) + i\theta_2(t)\right] \qquad (10.7)$$

Here, A_j and B_j represent the amplitude and width of the soliton respectively in the two cores. Also, κ_j and θ_j respectively represent the frequency and the phase of the soliton in the two cores for $j = 1, 2$, while g represents the functional form of the soliton. It depends on the type of nonlinearity in F. Finally, $x_j(t)$ represents the center position of the soliton in the two cores. For convenience, the following few integrals are defined that will be in use

$$I_{a,b,c}^{(l)} = \int_{-\infty}^{\infty} \tau_l^a g^b(\tau_l)\left(\frac{dg}{d\tau_l}\right)^c d\tau_l \qquad (10.8)$$

$$P_{a,b,c} = \int_{-\infty}^{\infty} x^a g^b\left[B_1(x - x_1(t))\right]g^c\left[B_2(x - x_2(t))\right]$$
$$\cos\left[(\kappa_1 - \kappa_2)x - (\kappa_1 x_1 - \kappa_2 x_2) - (\theta_1 - \theta_2)\right]dx \qquad (10.9)$$

$$Q_{a,b,c} = \int_{-\infty}^{\infty} x^a g^b\left[B_1(x - x_1(t))\right]g^c\left[B_2(x - x_2(t))\right]$$
$$\sin\left[(\kappa_1 - \kappa_2)x - (\kappa_1 x_1 - \kappa_2 x_2) - (\theta_1 - \theta_2)\right]dx \qquad (10.10)$$

for nonnegative integers a, b, and c and $l = 1, 2$. Here, $\tau_1 = B_1(t)(x - x_1(t))$ and $\tau_2 = B_2(t)(x - x_2(t))$. The integrals of motion in (10.3), (10.4), and (10.5), on using the soliton form in (10.6) and (10.7), respectively, reduce to [79]

$$E = \int_{-\infty}^{\infty}\left(|q|^2 + |r|^2\right)dx = \frac{A_1^2}{B_1}I_{0,2,0}^{(1)} + \frac{A_2^2}{B_2}I_{0,2,0}^{(2)} \qquad (10.11)$$

$$M = \frac{i}{2}\int_{-\infty}^{\infty}\{(qq_x^* - q^*q_x) + (rr_x^* - r^*r_x)\}dx = -\left(\kappa_1\frac{A_1^2}{B_1}I_{0,2,0}^{(1)} + \kappa_2\frac{A_2^2}{B_2}I_{0,2,0}^{(2)}\right) \qquad (10.12)$$

and

$$
\begin{aligned}
H &= \int_{-\infty}^{\infty} \left[\frac{1}{2} \left(|q_x|^2 + |r_x|^2 \right) - f\left(|q|^2\right) - f\left(|r|^2\right) - K\left(qr^* + rq^*\right) \right] dx \\
&= \frac{1}{2} \left(A_1^2 B_1 I_{0,0,2}^{(1)} + A_2^2 B_2 I_{0,0,2}^{(2)} \right) + \frac{1}{2} \left(\frac{\kappa_1^2 A_1^2}{B_1} I_{0,2,0}^{(1)} + \frac{\kappa_2^2 A_2^2}{B_2} I_{0,2,0}^{(2)} \right) \\
&\quad - 2K A_1 A_2 P_{0,1,1} - 2 \int_{-\infty}^{\infty} \int_0^I F(s) ds
\end{aligned}
\tag{10.13}
$$

The soliton parameters for the first core are now defined as

$$
A_1(t) = \left[\frac{I_{0,2,0}^{(1)} \int_{-\infty}^{\infty} |q|^4 \, dx}{I_{0,4,0}^{(1)} \int_{-\infty}^{\infty} |q|^2 \, dx} \right]^{\frac{1}{2}}
\tag{10.14}
$$

$$
B_1(t) = \left[\frac{I_{2,2,0}^{(1)} \int_{-\infty}^{\infty} |q|^2 \, dx}{I_{0,2,0}^{(1)} \int_{-\infty}^{\infty} x^2 |q|^2 \, dx} \right]^{\frac{1}{2}}
\tag{10.15}
$$

$$
\kappa_1(t) = \frac{i}{2} \frac{\int_{-\infty}^{\infty} (q q_x^* - q^* q_x) \, dx}{\int_{-\infty}^{\infty} |q|^2 dx}
\tag{10.16}
$$

and

$$
x_1(t) = \frac{\int_{-\infty}^{\infty} x |q|^2 dx}{\int_{-\infty}^{\infty} |q|^2 \, dx}
\tag{10.17}
$$

Treating the coupling terms as perturbation terms in (10.1) and differentiating these soliton parameters while using the soliton given by (10.6) as well as the modified parameter dynamics to obtain the evolution of the soliton parameters in presence of the coupling terms yields

$$
\frac{dE}{dt} = 0
\tag{10.18}
$$

$$
\frac{dA_1}{dt} = -2K A_2 B_1 \frac{Q_{0,3,1}}{I_{0,4,0}^{(1)}}
\tag{10.19}
$$

$$
\frac{dB_1}{dt} = 2K B_1^2 \frac{A_2}{A_1} \left(\frac{Q_{0,1,1}}{I_{0,2,0}^{(1)}} - \frac{Q_{0,3,1}}{I_{0,4,0}^{(1)}} \right)
\tag{10.20}
$$

$$
\frac{d\kappa_1}{dt} = 0
\tag{10.21}
$$

$$
\frac{dx_1}{dt} = -\kappa_1 + 2K \frac{B_1}{A_1^2} \frac{Q_{1,1,1}}{I_{0,2,0}^{(1)}}
\tag{10.22}
$$

$$\frac{d\theta_1}{dt} = -\frac{\kappa_1^2}{2} + K\frac{B_1 A_2}{A_1}\frac{P_{0,1,1}}{I_{0,2,0}^{(1)}} - 2K\frac{\kappa_1 B_1}{A_1^2}\frac{Q_{1,1,1}}{I_{0,2,0}^{(1)}}$$

$$+ \frac{1}{I_{0,2,0}^{(1)}}\int_{-\infty}^{\infty} F\big(A_1^2 g^2(\tau_1)\big) g^2(\tau_1) d\tau_1 \tag{10.23}$$

Similarly, the parameters for the solitons in the second core, as described by (10.2), are defined as

$$A_2(t) = \left[\frac{I_{0,2,0}^{(2)}\int_{-\infty}^{\infty}|r|^4 dx}{I_{0,4,0}^{(2)}\int_{-\infty}^{\infty}|r|^2 dx}\right]^{\frac{1}{2}} \tag{10.24}$$

$$B_2(t) = \left[\frac{I_{2,2,0}^{(2)}\int_{-\infty}^{\infty}|r|^2 dx}{I_{0,2,0}^{(2)}\int_{-\infty}^{\infty}x^2|r|^2 dx}\right]^{\frac{1}{2}} \tag{10.25}$$

$$\kappa_2(t) = \frac{i}{2}\frac{\int_{-\infty}^{\infty}(rr_x^* - r^*r_x)dx}{\int_{-\infty}^{\infty}|r|^2 dx} \tag{10.26}$$

$$x_2(t) = \frac{\int_{-\infty}^{\infty}x|r|^2 dx}{\int_{-\infty}^{\infty}|r|^2 dx} \tag{10.27}$$

These definitions also then lead to the following parameter dynamics for the solitons in twin-core couplers:

$$\frac{dA_2}{dt} = -2K A_1 B_2 \frac{Q_{0,3,1}}{I_{0,4,0}^{(2)}} \tag{10.28}$$

$$\frac{dB_2}{dt} = 2K B_2^2 \frac{A_1}{A_2}\left(\frac{Q_{0,1,1}}{I_{0,2,0}^{(2)}} - \frac{Q_{0,3,1}}{I_{0,4,0}^{(2)}}\right) \tag{10.29}$$

$$\frac{d\kappa_2}{dt} = 0 \tag{10.30}$$

$$\frac{dx_2}{dt} = -\kappa_2 + 2K\frac{B_2}{A_2^2}\frac{Q_{1,1,1}}{I_{0,2,0}^{(2)}} \tag{10.31}$$

$$\frac{d\theta_2}{dt} = -\frac{\kappa_2^2}{2} + K\frac{B_2 A_1}{A_2}\frac{P_{0,1,1}}{I_{0,2,0}^{(2)}} - 2K\frac{\kappa_2 B_2}{A_2^2}\frac{Q_{1,1,1}}{I_{0,2,0}^{(2)}}$$

$$+ \frac{1}{I_{0,2,0}^{(2)}}\int_{-\infty}^{\infty} F\big(A_2^2 g^2(\tau_2)\big) g^2(\tau_2) d\tau_2 \tag{10.32}$$

Here, (10.23) and (10.32) are obtained by differentiating (10.6) and (10.7) with respect to t and subtracting from its conjugate. Thus, from (10.18) one can see that the total energy in the two couplers remains a constant. However, the amplitude, width, frequency, center of mass, and phase of the solitons undergo a change as governed by (10.19)–(10.23) and (10.28)–(10.32), respectively. One can apply the new results developed in this section to obtain the parameter dynamics of the solitons in the cores with a particular kind of nonlinearity, namely F, and the corresponding form of the soliton g.

10.3   Multiple-Core Couplers

The arrays of active or passive coupled waveguides can be utilized for several practical applications. Multiple-core optical fibers are an area of current research interest. The use of these fibers for high-powered lasers is driving this interest. One possibility of applications would be all-optical switching between the cores. The equations derived in this chapter allow us to explore this possibility. This is one of the major reasons why multiple-core fibers are studied in optics. Multiple-core couplers can be arranged in rib or ring geometry as shown in Figures 10.3(a) and 10.3(b).

Figure 10.3(a) shows the rib geometry while Figure 10.3(b) shows the ring geometry for multiple-core fibers. In this section, (10.8), which is now valid for $1 \le l \le N$, shall be reused, and the following integrals will also be utilized

$$I_{a,b,c}^{(\mu,\nu)} = \int_{-\infty}^{\infty} x^a g^b [B_\mu(x - x_\mu(t))] g^c [B_\nu(x - x_\nu(t))]$$

$$\cos[(\kappa_\mu - \kappa_\nu)x - (\kappa_\mu x_\mu - \kappa_\nu x_\nu) - (\theta_\mu - \theta_\nu)] \, dx \qquad (10.33)$$

$$J_{a,b,c}^{(\mu,\nu)} = \int_{-\infty}^{\infty} x^a g^b [B_\mu(x - x_\mu(t))] g^c [B_\nu(x - x_\nu(t))]$$

$$\sin[(\kappa_\mu - \kappa_\nu)x - (\kappa_\mu x_\mu - \kappa_\nu x_\nu) - (\theta_\mu - \theta_\nu)] \, dx \qquad (10.34)$$

for $1 \le \mu, \nu \le N$ and nonnegative integers a, b, and c and $\tau_l = B_l(t)(x - x_l(t))$.

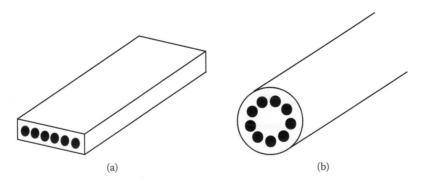

(a)   (b)

FIGURE 10.3
(a) and (b) Rib and ring geometry.

10.3.1 Coupling with Nearest Neighbors

This situation can be described by the N-coupled NLSE with the nearest neighbor linear coupling. They can be used as a more elaborate switching device. For multiple-core nonlinear fiber arrays, the governing equations for the ring geometry are given by [79]

$$iq_t^{(l)} + \frac{1}{2}q_{xx}^{(l)} + F(|q^{(l)}|^2)q^{(l)} = K(q^{(l-1)} + q^{(l+1)} - 2q^{(l)}) \qquad (10.35)$$

where $2 \leq l \leq N$ $q^{(0)} = q^{(N)}, q^{(N+1)} = q^{(1)}$. Here, once again, K represents the coupling coefficient between the neighboring cores and $q^{(l)}$ represents the optical field in the lth core. Also, (10.35) does not have infinitely many conservation laws. In this case, there are also as few as three integrals of motion. They are energy (E), linear momentum (M), and the Hamiltonian (H) that are respectively given by

$$E = \sum_{l=1}^{N} \int_{-\infty}^{\infty} |q^{(l)}|^2 dx \qquad (10.36)$$

$$M = \frac{i}{2} \sum_{l=1}^{N} \int_{-\infty}^{\infty} (q^{(l)}q_x^{(l)*} - q^{(l)*}q_x^{(l)}) dx \qquad (10.37)$$

and

$$H = \sum_{l=1}^{N} \int_{-\infty}^{\infty} \left[\frac{1}{2}|q_x^{(l)}|^2 - f(|q^{(l)}|^2) - K|q^{(l)} - q^{(l-1)}|^2 \right] dx \qquad (10.38)$$

Physically, E in (10.36) represents the sum of the individual energies in the N cores of the fiber. Since (10.35) is not solvable by the inverse scattering transform (IST) or any other method, it will be studied from a different perspective. For $K = 0$, the pulse in the lth core is written as

$$q^{(l)}(x, t) = A_l(t)g\left[B_l(t)(x - x_l(t))\right]\exp\left[-i\kappa_l(t)(x - x_l(t)) + i\theta_l(t)\right] \qquad (10.39)$$

where $1 \leq l \leq N$. Here also, just as in the case of twin-core couplers, A_l represents the amplitude of the soliton, B_l represents the width of the soliton, x_l is the center position, κ_l is the frequency, θ_l is the phase, and g represents the functional form of the soliton, which depends on the type of non–Kerr law nonlinearity. The notation $\tau_l = B_l(x - x_l(t))$ for $1 \leq l \leq N$ will be used. For such a general form of the soliton, the three integrals of motion are respectively given by

$$E = \sum_{l=1}^{N} \int_{-\infty}^{\infty} |q^{(l)}|^2 dx = \sum_{l=1}^{N} \frac{A_l^2}{B_l} I_{0,2,0}^{(l)} \qquad (10.40)$$

$$M = \frac{i}{2} \sum_{l=1}^{N} \int_{-\infty}^{\infty} (q^{(l)}q_x^{(l)*} - q^{(l)*}q_x^{(l)}) dx = -\sum_{l=1}^{N} \kappa_l \frac{A_l^2}{B_l} I_{0,2,0}^{(l)} \qquad (10.41)$$

and

$$H = \sum_{l=1}^{N} \int_{-\infty}^{\infty} \left[\frac{1}{2} |q_x^{(l)}|^2 - f(|q^{(l)}|^2) - K|q^{(l)} - q^{(l-1)}|^2 \right] dx$$

$$= \frac{1}{2} \sum_{l=1}^{N} \left[A_l^2 B_l J_{0,2,0}^{(l)} + \frac{\kappa_l^2 A_l^2}{B_l} I_{0,2,0}^{(l)} \right] - N \int_{-\infty}^{\infty} \int_{0}^{l} F(s) ds dx$$

$$- K \sum_{l=1}^{N} \left[\frac{A_l^2}{B_l} I_{0,2,0}^{(l)} + \frac{A_{l-1}^2}{B_{l-1}} I_{0,2,0}^{(l-1)} + 4 A_l A_{l-1} I_{0,1,1}^{(l,l-1)} \right] \tag{10.42}$$

For the case of multiple cores, the definitions of the soliton parameters are as follows:

$$A_l(t) = \left[\frac{I_{0,2,0}^{(l)} \int_{-\infty}^{\infty} |q^{(l)}|^4 dx}{I_{0,4,0}^{(l)} \int_{-\infty}^{\infty} |q^{(l)}|^2 dx} \right]^{\frac{1}{2}} \tag{10.43}$$

$$B_l(t) = \left[\frac{I_{2,2,0}^{(l)} \int_{-\infty}^{\infty} |q^{(l)}|^2 dx}{I_{0,2,0}^{(l)} \int_{-\infty}^{\infty} x^2 |q^{(l)}|^2 dx} \right]^{\frac{1}{2}} \tag{10.44}$$

$$\kappa_l(t) = \frac{i}{2} \frac{\int_{-\infty}^{\infty} (q^{(l)} q_x^{(l)*} - q^{(l)*} q_x^{(l)}) \, dx}{\int_{-\infty}^{\infty} |q^{(l)}|^2 \, dx} \tag{10.45}$$

and

$$x_l(t) = \frac{\int_{-\infty}^{\infty} x |q^{(l)}|^2 \, dx}{\int_{-\infty}^{\infty} |q^{(l)}|^2 \, dx} \tag{10.46}$$

These definitions are valid for $1 \leq l \leq N$. Now, differentiating these parameters with respect to t and using (10.39) while treating the coupling terms as perturbation terms, the following evolution equations for the soliton parameters in multiple cores are obtained

$$\frac{dE}{dt} = 0 \tag{10.47}$$

$$\frac{dA_l}{dt} = -\frac{2K B_l}{I_{0,4,0}^{(l)}} \left[A_{l-1} J_{0,3,1}^{(l,l-1)} + A_{l+1} J_{0,3,1}^{(l,l+1)} \right] \tag{10.48}$$

$$\frac{dB_l}{dt} = \frac{2K B_l^2}{A_l} \frac{1}{I_{0,4,0}^{(l)}} \left[\left\{ A_{l-1} J_{0,1,1}^{(l,l-1)} + A_{l+1} J_{0,1,1}^{(l,l+1)} \right\} \right.$$

$$\left. - 2 \left\{ A_{l-1} J_{0,3,1}^{(l,l-1)} + A_{l+1} J_{0,3,1}^{(l,l+1)} \right\} \right] \tag{10.49}$$

$$\frac{d\kappa_l}{dt} = 0 \tag{10.50}$$

$$\frac{dx_l}{dt} = -\kappa_1 - \frac{2K}{I_{0,2,0}^{(l)}} \frac{B_l}{A_l} \left(A_{l-1} J_{1,1,1}^{(l,l-1)} + A_{l+1} J_{1,1,1}^{(l,l+1)} \right) \tag{10.51}$$

$$\frac{d\theta_l}{dt} = -\frac{\kappa_1^2}{2} + \frac{I_{0,0,2}^{(l)}}{I_{0,2,0}^{(l)}} \frac{B_l^2}{2} - \frac{1}{I_{0,2,0}^{(l)}} \int_{-\infty}^{\infty} F\left(A_l^2 g^2(\tau_l) \right) g^2(\tau_l) \, d\tau_l$$

$$+ \frac{2K}{I_{0,2,0}^{(l)}} \frac{\kappa_l B_l}{A_l} \left(A_{l-1} J_{1,1,1}^{(l,l-1)} + A_{l+1} J_{1,1,1}^{(l,l+1)} \right)$$

$$- \frac{K}{I_{0,2,0}^{(l)}} \frac{B_l}{A_l} \left(A_{l-1} I_{0,1,1}^{(l,l-1)} + A_{l+1} I_{0,1,1}^{(l,l+1)} \right) \tag{10.52}$$

where $1 \le l \le N$. From (10.47), one can conclude that the total energy in the couplers remains a constant. However, the amplitude, width, frequency, center of mass, and phase of the solitons in the couplers experience a change as governed by (10.48)–(10.52). If, however, the law of coupling with the nearest neighbor is

$$iq_t^{(l)} + \frac{1}{2} q_{xx}^{(l)} + F(|q^{(l)}|^2) q^{(l)} = K \left(q^{(l-1)} + q^{(l+1)} \right) \tag{10.53}$$

the same parameter dynamics appear and other features also stay the same. However, the Hamiltonian, in this case, is given by

$$H = \sum_{l=1}^{N} \int_{-\infty}^{\infty} \left[\frac{1}{2} |q_x^{(l)}|^2 - f\left(|q^{(l)}|^2 \right) \right.$$

$$\left. - K \left(q^{(l)} q^{(l-1)*} + q^{(l)*} q^{(l-1)} \right) - K \left(q^{(l)} q^{(l+1)*} + q^{(l)*} q^{(l+1)} \right) \right] dx$$

$$= \frac{1}{2} \sum_{l=1}^{N} A_l^2 B_l I_{0,0,2}^{(l)} + \frac{1}{2} \sum_{l=1}^{N} \frac{\kappa_l^2 A_l^2}{B_l} I_{0,0,2}^{(l)} - N \int_{-\infty}^{\infty} \int_0^I F(s) ds dx$$

$$+ 2K \sum_{l=1}^{N} A_l \left\{ A_{l-1} I_{0,1,1}^{(l,l-1)} + A_{l+1} I_{0,1,1}^{(l,l+1)} \right\} \tag{10.54}$$

10.3.2 Coupling with All Neighbors

For pulses propagating through N coupled nonlinear fiber arrays, when coupling with all neighbors, the equation is given in the following dimensionless form

$$iq_t^{(l)} + \frac{1}{2} q_{xx}^{(l)} + F\left(|q^{(l)}|^2 \right) q^{(l)} = \sum_{n=1}^{N} \lambda_{lng}^{(n)} \tag{10.55}$$

where $1 \le l \le N$. In (10.55), the right-hand side is due to the linear coupling term between the fiber arrays and λ_{ij} is the linear coupling coefficient for $1 \le i, j \le N$. The coupling coefficients form a symmetric matrix Λ of N rows and N columns. So, $\Lambda = (\lambda_{ij})_{N \times N}$ where $\lambda_{ii} = 0$ for $1 \le i \le N$ and $\lambda_{ij} = \lambda_{ji} \, \forall \, i, j$.

For example, in a twin-core fiber array

$$\Lambda = \begin{bmatrix} 0 & \lambda_{12} \\ \lambda_{12} & 0 \end{bmatrix} \tag{10.56}$$

while for a triple-core fiber array

$$\Lambda = \begin{bmatrix} 0 & \lambda_{12} & \lambda_{13} \\ \lambda_{12} & 0 & \lambda_{23} \\ \lambda_{13} & \lambda_{23} & 0 \end{bmatrix} \tag{10.57}$$

and so on. The functional differential-difference equation (10.55) is not integrable by the IST due to the presence of the coupling coefficients and because of the arbitrary function F. For $\lambda_{ln} = 0$, the pulse in the lth core is given by (10.39). Moreover, (10.55) does not have infinitely many conserved quantities. In fact, in this case also, it has as few as three integrals of motion. They are energy (E), linear momentum (M), and the Hamiltonian (H). Although the energy and momentum stay the same as in the previous subsection, the Hamiltonian here is given by

$$H = \sum_{l=1}^{N} \int_{-\infty}^{\infty} \left[\frac{1}{2} |q_x^{(l)}|^2 - f\left(|q^{(l)}|^2\right) - \sum_{n=1}^{N} \lambda_{ln} \left(q^{(n)} q^{(n-1)*} - q^{(n)*} q^{(n-1)} \right) \right] dx$$

$$= \frac{1}{2} \sum_{l=1}^{N} \left[A_l^2 B_l J_{0,2,0}^{(l)} + \frac{\kappa_l^2 A_l^2}{B_l} I_{0,2,0}^{(l)} \right] - 2 \sum_{l=1}^{N} \sum_{n=1}^{N} \lambda_{ln} A_n A_{n-1} I_{0,1,1}^{(n,n-1)}$$

$$- N \int_{-\infty}^{\infty} \int_{0}^{I} F(s) ds dx \tag{10.58}$$

where $q^{(0)} = q^{(N)}$. The soliton parameters are defined as in (10.43)–(10.46). Again, differentiating these parameters with respect to t and using (10.39) while treating the coupling terms as perturbation terms, the following evolution equations for the soliton parameters in multiple cores are obtained

$$\frac{dE}{dt} = 2 \sum_{l=1}^{N} \sum_{n=1}^{N} A_l A_n J_{0,1,1}^{(l,n)} \tag{10.59}$$

$$\frac{dA_l}{dt} = \frac{2B_l}{I_{0,4,0}^{(l)}} \sum_{n=1}^{N} \lambda_{ln} A_n J_{0,3,1}^{(n,l)} \tag{10.60}$$

$$\frac{dB_l}{dt} = \frac{2}{A_l} \frac{1}{I_{0,2,0}^{(l)}} \sum_{n=1}^{N} \lambda_{ln} A_n J_{0,1,1}^{(n,l)} \tag{10.61}$$

$$\frac{d\kappa_l}{dt} = 0 \tag{10.62}$$

$$\frac{dx_l}{dt} = -\kappa_1 + \frac{2}{I_{0,2,0}^{(l)}} \frac{B_l}{A_l} \sum_{n=1}^{N} \lambda_{ln} A_n I_{1,1,1}^{(n,l)} \tag{10.63}$$

$$\frac{d\theta_l}{dt} = -\frac{\kappa_1^2}{2} - \frac{J_{0,2,0}^{(l)}}{I_{0,2,0}} \frac{B_l^2}{2} - \frac{1}{I_{0,2}^{(l)}} \frac{B_l}{A_l} \sum_{n=1}^{N} \lambda_{ln} A_n I_{0,1,1}^{(l,n)}$$

$$+ \frac{1}{I_{0,2,0}^{(l)}} \int_{-\infty}^{\infty} F\left(A_l^2 g^2(\tau_l)\right) g^2(\tau_l) d\tau_l \tag{10.64}$$

where $1 \leq l \leq N$. Thus, from (10.59)–(10.64), it is seen that the energy, amplitude, width, and phase of the solitons in each array changes in the presence of the coupling terms; however, the frequency in each array stays constant.

10.4 Magneto-Optic Waveguides

Optical waveguides, which are made from thin garnet films, have been of interest since they were first demonstrated in 1972. In addition, it is often argued that the existence of magnetostatic waves implies devices that have considerable advantages over acousto-optic applications since garnet films operate well into the high gigahertz frequency range. This is an important microwave frequency range, and garnet films offer the added flexibility of magnetic tunability. Indeed, a whole range of microwave signal processing devices, such as filters, correlators, spectrum analyzer switches, modulator frequency shifters, and tunable filters, are either in use or appear to be on the horizon. Such devices will be even more useful if power can be added in as another degree of flexibility. In other words, the study of nonlinear magneto-optical interactions is of prime importance. It is now easy to make the fundamental integrated optical building block, which is the channel waveguide. It is also possible to envision the future integration of magneto-optic devices with semiconducting substrates containing active devices, such as laser detectors and amplifiers. It would appear, then, that there has never been a better time to pursue the combination of nonlinearity and magneto-optics. This is true even though magneto-optics have recently been described as a "stepchild" of integrated optics. Some of this impression originates from a desire to "insert" magneto-optics into known designs rather than address and control the fascinating complexity of the materials. The manner in which an external magnetic field, applied to a waveguide containing third-order optically nonlinear material ($\chi^{(3)}$) and magneto-optical elements can force bright solitons from a state of attraction to a state of isolation from each other [84].

Until recently, any discussion on envelope solitons in nonlinear optics was based upon the usage of third-order nonlinearity. The general form of the polarization induced by a relatively high-power electromagnetic wave, during its passage through a dielectric material, however apart from its obvious dependence on $\chi^{(1)}$, depends also on $\chi^{(2)}$ as well as $\chi^{(3)}$. Truncation of polarization at the third order is appropriate for most materials and, furthermore, many elements of these tensors are usually zero. This means that tensors are often reducible to a single, independent parameter or, at most, a small number of independent ones, by application of crystal symmetry operations. Indeed the materials needed for real applications are often isotropic amorphous or possess a rather high crystal symmetry. In an $\chi^{(2)}$ material, however, two field components can mix together to produce a third one that is, once again, at the fundamental frequency—in other words, back-mixing (cascading) occurs involving second-harmonic waves and the complex conjugate of the fundamental wave. Furthermore, experimental evidence shows that the back-mixing process is clearly observable, even for large linear phase mismatching.

10.4.1 Mathematical Analysis

The dimensionless form of the vector NLSE in a magneto-optic waveguide, due to the polarization control by the magnetic field, is given by [84]

$$iu_t + \frac{1}{2}u_{xx} + \{F(|u|^2) + \alpha F(|v|^2)\}u = Qv \tag{10.65}$$

$$iv_t + \frac{1}{2}v_{xx} + \{F(|v|^2) + \alpha F(|u|^2)\}v = Qu \tag{10.66}$$

Here $Q = Q(x)$ is called the *magneto-optical parameter* and α is a constant. These equations are derived for the case of Kerr law nonlinearity in a $\chi^{(3)}$ medium [298]. Equations (10.65) and (10.66) are not integrable. Therefore, these equations will be studied by the aid of soliton perturbation theory. Assume that the solitons in the two waveguides are respectively given by (10.6) and (10.7). For convenience, the following integrals are defined

$$G_{l,m,n} = \int_{-\infty}^{\infty} Q(x)x^l g^m(\tau_1)g^n(\tau_2)\sin\{\kappa_1(x-x_1) - \kappa_2(x-x_2) + (\theta_1 - \theta_2)\}dx$$

$$\tag{10.67}$$

$$H_{l,m,n} = \int_{-\infty}^{\infty} Q(x)x^l g^m(\tau_1)g^n(\tau_2)\cos\{\kappa_1(x-x_1) - \kappa_2(x-x_2) + (\theta_1 - \theta_2)\}dx$$

$$\tag{10.68}$$

for nonnegative integers l, m, and n, while $k = 1, 2$, where $\tau_k = B_k(t)(x - x_k(t))$. For such general forms of the solitons given by (10.6) and (10.7), the energy is given by (10.11). With the same definition of the soliton parameters as in (10.14)–(10.17), the parameter dynamics, on treating the cross-phase modulation term and the magneto-optical parameter term as perturbation terms, are given by

$$\frac{d\Gamma}{dt} = 0 \tag{10.69}$$

$$\frac{dA_1}{dt} = 2A_2 B_1 \frac{G_{0,3,1}}{I_{0,4,0}^{(1)}} \tag{10.70}$$

$$\frac{dB_1}{dt} = 2\frac{A_2}{A_1} \frac{G_{0,1,1}}{I_{0,2,0}^{(1)}} \tag{10.71}$$

$$\frac{d\kappa_1}{dt} = \frac{2\alpha}{I_{0,2,0}^{(1)}} \int_{-\infty}^{\infty} g^2(\tau_1) g(\tau_2) g'(\tau_2) F\left(A_{2}^2 g^2(\tau_2)\right) dx \tag{10.72}$$

$$\frac{dx_1}{dt} = 2\frac{A_2 B_1}{A_1} \frac{G_{1,1,1}}{I_{0,2,0}^{(1)}} \tag{10.73}$$

and

$$\frac{d\theta_1}{dt} = -\frac{\kappa_1^2}{2} - 2\kappa_1 \frac{A_2 B_1}{A_1} \frac{G_{1,1,1}}{I_{0,2,0}^{(1)}} + \alpha F\left(A_{2}^2 g^2(\tau_2)\right)$$
$$+ \frac{1}{I_{0,2,0}^{(1)}} \int_{-\infty}^{\infty} g^2(\tau_1) F\left(A_{1}^2 g^2(\tau_1)\right) d\tau_1 - \frac{A_2 B_1}{A_1} \frac{H_{0,1,1}}{I_{0,2,0}^{(1)}} \tag{10.74}$$

Similarly, using the definitions given by (10.24)–(10.27), the evolution equations of the soliton parameters are

$$\frac{dA_2}{dt} = 2A_1 B_2 \frac{G_{0,3,1}}{I_{0,4,0}^{(2)}} \tag{10.75}$$

$$\frac{dB_2}{dt} = 2\frac{A_1}{A_2} \frac{G_{0,1,1}}{I_{0,2,0}^{(2)}} \tag{10.76}$$

$$\frac{d\kappa_2}{dt} = \frac{2\alpha}{I_{0,2,0}^{(2)}} \int_{-\infty}^{\infty} g^2(\tau_2) g(\tau_1) g'(\tau_1) F\left(A_{1}^2 g^2(\tau_1)\right) dx \tag{10.77}$$

$$\frac{dx_2}{dt} = 2\frac{A_1 B_2}{A_2} \frac{G_{1,1,1}}{I_{0,2,0}^{(2)}} \tag{10.78}$$

and

$$\frac{d\theta_2}{dt} = -\frac{\kappa_2^2}{2} - 2\kappa_2 \frac{A_1 B_2}{A_2} \frac{G_{1,1,1}}{I_{0,2,0}^{(2)}} + \alpha F\left(A_{1}^2 g^2(\tau_1)\right)$$
$$+ \frac{1}{I_{0,2,0}^{(2)}} \int_{-\infty}^{\infty} g^2(\tau_2) F\left(A_{2}^2 g^2(\tau_2)\right) d\tau_2 - \frac{A_1 B_2}{A_2} \frac{H_{0,1,1}}{I_{0,2,0}^{(2)}} \tag{10.79}$$

Thus, from (10.69), the energy of the soliton remains constant as it is a conserved quantity for the set of equations in (10.65) and (10.66). However, the amplitude, width, frequency, center of mass, and phase of the solitons undergo a change as governed by (10.70)–(10.74) and (10.75)–(10.79), respectively. These are the parameter dynamics for the solitons due to non–Kerr law nonlinearity in a magneto-optic waveguide.

Exercises

1. For twin-core couplers, prove that the three integrals of motion given by (10.3), (10.4), and (10.5) respectively reduce to (10.11), (10.12), and (10.13) for the pulses given by (10.6) and (10.7) in the two couplers.

2. For multiple-core couplers, with nearest neighbor coupling, prove that the three integrals of motion given by (10.36), (10.37), and (10.38) respectively reduce to (10.40), (10.41), and (10.42) for pulse in the lth core given by (10.39).

3. For multiple-core couplers, with all neighbors coupling, prove that the Hamiltonian is given by (10.58).

11

Optical Bullets

This chapter gives a detailed overview of the theory of optical bullets that is due the nonlinear Schrödinger's equation (NLSE) in multidimensions. Section 11.1 presents an introduction to optical bullets, including their physics. Section 11.2 deals with the mathematical issues of optical bullets, including the structure of the NLSE and its conserved quantities.

11.1 Introduction

First conceived by Silberberg in 1990, an optical bullet is an optical pulse localized in space and time that maintains its spatiotemporal shape while propagating. Formation of optical bullets requires an anomalous dispersive medium where it is possible to achieve exact balance of diffraction and dispersion with optical nonlinearity. Dispersion and diffraction attempt to spread the pulse in longitudinal and transverse directions, respectively. Conversely, self-focusing effects attempt to compress the pulse. Consequently, a self-trapped optical soliton could be formed when these opposing forces balance each other. It has been realized that the solitary wave solutions of the cubic NLSE are unstable in more than one transverse dimension, hence standard nonlinear optical materials wherein the refractive index is strictly proportional to the intensity of light do not allow stable light bullets. Therefore, all theoretical and experimental investigations have been carried out in media that deviate from Kerr nonlinearity; particularly it has been shown theoretically that optical bullets are stable in saturable media. The mathematical model that can describe an optical bullet is either the two- or three-dimensional modified NLSE (mNLSE), depending on whether diffraction is limited to one or two transverse dimensions. Although the mNLSE permits the propagation of stable light bullets, these are not true solitons in a strict mathematical sense. The reason is that, unlike $1 + 1$ dimension solitons in Kerr media, which survive collisions with no loss of energy, these light bullets are found to possess less energy after collisions. Moreover, in favorable situations, they may lead to a new phenomenon, such as soliton fusion, fission, soliton tunneling, and formation of spiraling light bullets. Nevertheless, in physics literature and despite protestation of a few pursuits, the pragmatic use of the term solitons for these robust pulses has become common.

11.2 1 + 3 Dimensions

The dynamics of optical bullets are governed by the NLSE in $1+3$ dimensions. For general non-Kerr law, the NLSE in $1+3$ dimensions in the dimensionless form is given by

$$iq_t + \frac{1}{2}\nabla^2 q + F(|q|^2)q = 0 \tag{11.1}$$

where

$$\nabla^2 \equiv \frac{\partial^2}{\partial x^2} + \frac{\partial^2}{\partial y^2} + \frac{\partial^2}{\partial z^2}$$

and F is a real valued algebraic function. In general, (11.1) is not integrable. This nonintegrability is not necessarily related to the nonlinear term in (11.1).

11.2.1 Integrals of Motion

Equation (11.1) has several symmetries in $1 + 2$ dimensions, including rotation, dilatation, the Galelian transformations, and the Talanov lens transformations. On using the symmetry reductions, the exact solutions to (11.1) in $1 + 2$ dimensions were found for Kerr law case, in terms of Jacobi's elliptic functions. In addition, there also exists exact solutions for the self-focusing and the self-defocusing cases that are multivalued at each point of real space. Moreover, nonstationary singular solutions were also obtained to (11.1). In fact, for (11.1) there exists at least four integrals of motion. They are

$$E = \int_{-\infty}^{\infty} \int_{-\infty}^{\infty} \int_{-\infty}^{\infty} |q|^2 \, dx\,dy\,dz \tag{11.2}$$

$$P = \frac{i}{2} \int_{-\infty}^{\infty} \int_{-\infty}^{\infty} \int_{-\infty}^{\infty} (q\nabla q^* - q^*\nabla q)dx\,dy\,dz \tag{11.3}$$

$$M = \frac{i}{2} \int_{-\infty}^{\infty} \int_{-\infty}^{\infty} \int_{-\infty}^{\infty} r \times (q\nabla q^* - q^*\nabla q)dx\,dy\,dz \tag{11.4}$$

and

$$H = \int_{-\infty}^{\infty} \int_{-\infty}^{\infty} \int_{-\infty}^{\infty} \left[\frac{1}{2}(\nabla q) \cdot (\nabla q^*) - f(|q|^2)\right]dx\,dy\,dz \tag{11.5}$$

In (11.4), $\mathbf{r}$ represents the position vector, namely $\mathbf{r} = (x, y, z)$. The integrals of motion are respectively known as energy (E), linear momentum ($\mathbf{\Gamma}$), angular momentum ($\mathbf{M}$), and the Hamiltonian (H). Also, the conserved quantity related to the Talanov lens transformation was found in 1985. It is important to remark here that in the special case of power law nonlinearity, an additional conserved quantity exists that is given by [361]

$$C = \int_{-\infty}^{\infty} \int_{-\infty}^{\infty} \int_{-\infty}^{\infty} \left(|rq + 2it\nabla q|^2 - \frac{4t^2}{p+1}|q|^{2p+2} \right) dx\, dy\, dz \qquad (11.6)$$

In fact, it was first pointed out by Kuznetsov and Turitsyn in 1985 that the invariant (C) in (11.6) is a consequence of Noether's theorem.

Since there is no inverse scattering solution to (11.1), the dynamics of parameters of an optical bullet will be derived from their corresponding definitions. For this, it will be assumed that the solution of (11.1), although not integrable, is given in the form [78]

$$q(x, y, z; t) = A(t)g[B_1(t)\{x - \bar{x}(t)\}, B_2(t)\{y - \bar{y}(t)\}, B_3(t)\{z - \bar{z}(t)\}]$$

$$\exp[-i\kappa_1(t)\{x - \bar{x}(t)\} - i\kappa_2(t)\{y - \bar{y}(t)\} - i\kappa_3(t)\{z - \bar{z}(t)\} + i\theta(t)] \qquad (11.7)$$

where g represents the functional form of the shape described by the NLSE and depends on the type of nonlinearity in (11.1). Also in (11.7), $A(t)$ represents the amplitude, while $B_j(t)$ for $j = 1, 2, 3$ represents the width in the x, y, and z directions, respectively. Then, $\kappa_j(t)$ ($j = 1, 2, 3$) as the frequency of the soliton in the x, y, and z directions, respectively. Finally, $(\bar{x}(t), \bar{y}(t), \bar{z}(t))$ is the coordinate of the center of mass and $\theta(t)$ represents the phase. For convenience, the following integral is defined

$$I_{a,b,c,p}^{\alpha,\beta,\gamma} = \int_{-\infty}^{\infty} \int_{-\infty}^{\infty} \int_{-\infty}^{\infty} (\tau_1)^a (\tau_2)^b (\tau_3)^c g^p(\tau_1, \tau_2, \tau_3)$$

$$\cdot \left(\frac{\partial g}{\partial \tau_1} \right)^\alpha \left(\frac{\partial g}{\partial \tau_2} \right)^\beta \left(\frac{\partial g}{\partial \tau_3} \right)^\gamma d\tau_1 d\tau_2 d\tau_3 \qquad (11.8)$$

with nonnegative integers $a, b, c, p, \alpha, \beta$, and γ where $\tau_1 = B_1(t)(x - \bar{x}(t))$, $\tau_2 = B_2(t)(y - \bar{y}(t))$ and $\tau_3 = B_3(t)(z - \bar{z}(t))$. For such a general form of the bullet, given by (11.7), the integrals of motion from (11.2), (11.3), (11.4), and (11.5) respectively simplify to

$$E = \int_{-\infty}^{\infty} \int_{-\infty}^{\infty} \int_{-\infty}^{\infty} |q|^2\, dx\, dy\, dz = \frac{A^2}{B_1 B_2 B_3} I_{0,0,0,2}^{0,0,0} \qquad (11.9)$$

$$P = \frac{i}{2} \int_{-\infty}^{\infty} \int_{-\infty}^{\infty} \int_{-\infty}^{\infty} (q\nabla q^* - q^*\nabla q)\, dx\, dy\, dz = -\frac{A^2}{B_1 B_2 B_3} I_{0,0,0,2}^{0,0,0}(\kappa_1, \kappa_2, \kappa_3)$$

$$(11.10)$$

$$M = \frac{i}{2} \int_{-\infty}^{\infty} \int_{-\infty}^{\infty} \int_{-\infty}^{\infty} r \times (q \nabla q^* - q^* \nabla q) \, dx \, dy \, dz$$

$$= \frac{A^2}{B_1 B_2 B_3} \left(\frac{\kappa_2}{B_3} I_{0,0,1,2}^{0,0,0} - \frac{\kappa_3}{B_2} I_{0,1,0,2}^{0,0,0}, \right.$$

$$\left. \frac{\kappa_3}{B_1} I_{1,0,0,2}^{0,0,0} - \frac{\kappa_1}{B_3} I_{0,0,1,2}^{0,0,0}, \frac{\kappa_1}{B_2} I_{0,1,0,2}^{0,0,0} - \frac{\kappa_2}{B_1} I_{1,0,0,2}^{0,0,0} \right)$$

$$\tag{11.11}$$

and

$$H = \int_{-\infty}^{\infty} \int_{-\infty}^{\infty} \int_{-\infty}^{\infty} \left[\frac{1}{2} (\nabla q) \cdot (\nabla q^*) - f(|q|^2) \right] dx \, dy \, dz$$

$$= \frac{A^2}{2 B_1 B_2 B_3} \left[B_1^2 I_{0,0,0,0}^{2,0,0} + B_2^2 I_{0,0,0,0}^{0,2,0} + B_3^2 I_{0,0,0,0}^{0,0,2} + (\kappa_1^2 + \kappa_2^2 + \kappa_3^2) I_{0,0,0,2}^{0,0,0} \right]$$

$$- \int_{-\infty}^{\infty} \int_{-\infty}^{\infty} \int_{-\infty}^{\infty} \int_0^I f(s) \, ds \, dx \, dy \, dz \tag{11.12}$$

where the intensity I is given by $|q|^2$.

11.2.2   Parameter Evolution

There are ten parameters for the case of optical bullets. They are the amplitude $A(t)$, the width $B_j(t)$ ($j = 1, 2, 3$) in the x, y, and z directions, the frequencies $\kappa_j(t)$ in the three directions, and the coordinates of the center of mass of the soliton that is given by $(\bar{x}, \bar{y}, \bar{z})$. These parameters are defined as follows

$$A(t) = \left[\frac{I_{0,0,0}^{0,0,0} \int_{-\infty}^{\infty} \int_{-\infty}^{\infty} \int_{-\infty}^{\infty} |q|^4 \, dx \, dy \, dz}{I_{0,0,0,4}^{0,0,0} \int_{-\infty}^{\infty} \int_{-\infty}^{\infty} \int_{-\infty}^{\infty} |q|^2 \, dx \, dy \, dz} \right]^{\frac{1}{2}} \tag{11.13}$$

The definition of the width of the bullet in the x, y, and z directions are respectively defined as

$$B_1(t) = \left[\frac{I_{2,0,0,2}^{0,0,0} \int_{-\infty}^{\infty} \int_{-\infty}^{\infty} \int_{-\infty}^{\infty} |q|^2 \, dx \, dy \, dz}{I_{0,0,0,2}^{0,0,0} \int_{-\infty}^{\infty} \int_{-\infty}^{\infty} \int_{-\infty}^{\infty} \tau_1^2 |q|^2 \, dx \, dy \, dz} \right]^{\frac{1}{2}} \tag{11.14}$$

$$B_2(t) = \left[\frac{I_{0,2,0,2}^{0,0,0} \int_{-\infty}^{\infty} \int_{-\infty}^{\infty} \int_{-\infty}^{\infty} |q|^2 \, dx \, dy \, dz}{I_{0,0,0,2}^{0,0,0} \int_{-\infty}^{\infty} \int_{-\infty}^{\infty} \int_{-\infty}^{\infty} \tau_2^2 |q|^2 \, dx \, dy \, dz} \right]^{\frac{1}{2}} \tag{11.15}$$

and

$$B_3(t) = \left[\frac{I_{0,0,2,2}^{0,0,0} \int_{-\infty}^{\infty} \int_{-\infty}^{\infty} \int_{-\infty}^{\infty} |q|^2 \, dx \, dy \, dz}{I_{0,0,0,2}^{0,0,0} \int_{-\infty}^{\infty} \int_{-\infty}^{\infty} \int_{-\infty}^{\infty} \tau_3^2 |q|^2 \, dx \, dy \, dz} \right]^{\frac{1}{2}} \tag{11.16}$$

The frequency components in the x, y, and z directions are respectively defined as

$$\kappa_1(t) = \frac{i}{2} \frac{\int_{-\infty}^{\infty}\int_{-\infty}^{\infty}\int_{-\infty}^{\infty}(qq_x^* - q^*q_x)\,dx\,dy\,dz}{\int_{-\infty}^{\infty}\int_{-\infty}^{\infty}\int_{-\infty}^{\infty}|q|^2\,dx\,dy\,dz} \qquad (11.17)$$

$$\kappa_2(t) = \frac{i}{2} \frac{\int_{-\infty}^{\infty}\int_{-\infty}^{\infty}\int_{-\infty}^{\infty}(qq_y^* - q^*q_y)\,dx\,dy\,dz}{\int_{-\infty}^{\infty}\int_{-\infty}^{\infty}\int_{-\infty}^{\infty}|q|^2\,dx\,dy\,dz} \qquad (11.18)$$

and

$$\kappa_3(t) = \frac{i}{2} \frac{\int_{-\infty}^{\infty}\int_{-\infty}^{\infty}\int_{-\infty}^{\infty}(qq_z^* - q^*q_z)\,dx\,dy\,dz}{\int_{-\infty}^{\infty}\int_{-\infty}^{\infty}\int_{-\infty}^{\infty}|q|^2\,dx\,dy\,dz} \qquad (11.19)$$

Finally, the coordinates of the center of mass of the soliton $(\bar{x}, \bar{y}, \bar{z})$ are defined as

$$\bar{x}(t) = \frac{\int_{-\infty}^{\infty}\int_{-\infty}^{\infty}\int_{-\infty}^{\infty} x|q|^2 dx\,dy\,dz}{\int_{-\infty}^{\infty}\int_{-\infty}^{\infty}\int_{-\infty}^{\infty} |q|^2 dx\,dy\,dz} \qquad (11.20)$$

$$\bar{y}(t) = \frac{\int_{-\infty}^{\infty}\int_{-\infty}^{\infty}\int_{-\infty}^{\infty} y|q|^2 dx\,dy\,dz}{\int_{-\infty}^{\infty}\int_{-\infty}^{\infty}\int_{-\infty}^{\infty} |q|^2 dx\,dy\,dz} \qquad (11.21)$$

and

$$\bar{z}(t) = \frac{\int_{-\infty}^{\infty}\int_{-\infty}^{\infty}\int_{-\infty}^{\infty} z|q|^2 dx\,dy\,dz}{\int_{-\infty}^{\infty}\int_{-\infty}^{\infty}\int_{-\infty}^{\infty} |q|^2 dx\,dy\,dz} \qquad (11.22)$$

Now, differentiating these parameters with respect to t and using (11.7), the following evolution equations for the soliton parameters are obtained

$$\frac{dE}{dt} = 0 \qquad (11.23)$$

$$\frac{dA}{dt} = 0 \qquad (11.24)$$

$$\frac{dB_1}{dt} = \frac{dB_2}{dt} = \frac{dB_3}{dt} = 0 \qquad (11.25)$$

$$\frac{d\kappa_1}{dt} = \frac{d\kappa_2}{dt} = \frac{d\kappa_3}{dt} = 0 \qquad (11.26)$$

$$\frac{d\bar{x}}{dt} = -\kappa_1 \qquad (11.27)$$

$$\frac{d\bar{y}}{dt} = -\kappa_2 \qquad (11.28)$$

and

$$\frac{d\bar{z}}{dt} = -\kappa_3 \qquad (11.29)$$

Again, differentiating q given by (11.7), partially with respect to t, and subtracting from its conjugate while utilizing (11.1) yields the evolution of the phase of the optical bullet that is given by

$$\frac{d\theta}{dt} = -\frac{\kappa_1^2 + \kappa_2^2 + \kappa_3^2}{2} + \frac{1}{2I_{0,0,2}^{0,0,0}}\left(B_1^2 I_{0,0,0,0}^{2,0,0} + B_2^2 I_{0,0,0,0}^{0,2,0} + B_3^2 I_{0,0,0,0}^{0,0,2}\right) - \frac{1}{I_{0,0,2}^{0,0,0}}$$

$$\cdot \int_{-\infty}^{\infty} \int_{-\infty}^{\infty} \int_{-\infty}^{\infty} g^2(\tau_1, \tau_2, \tau_3) F\left(A^2 g^2(\tau_1, \tau_2, \tau_3)\right) d\tau_1 d\tau_2 d\tau_3 \qquad (11.30)$$

Thus, from (11.23)–(11.26), the energy, amplitude, width, and frequency of the bullet remain constant. However, the center of mass and phase of the bullet undergo a change as governed by (11.27)–(11.30).

12

Epilogue

The history of solitons is very long. It began one day in August 1834, when Scottish engineer John Scott Russell first observed, albeit accidentally, a large solitary water wave in a canal near Edinburgh. This unusual solitary wave, which is a localized large mass of water, traveled more than 2 km without noticeable decay in height and change of shape. Ten years later, Russell reported this unusual observation to the British Association for the Advancement of Science. The contemporary scientific community did not take note of it until 1895, when Dutch mathematicians Korteweg and de Vries theoretically proved through the Korteweg–de Vries (KdV) equation that solitary waves were indeed possible. Seven decades then passed, without anything memorable being reported. Meanwhile, Martin Kruskal and Normal Zabusky extended the investigation and, in 1965, came out with the surprising result that interaction of two solitary waves is identical to that of two colliding elementary particles. Thus, they coined the name *soliton* for such waves. Soon afterward, using inverse scattering transform theory, Zakharov and Shabat of the Soviet Union showed that the nonlinear Schrödinger's equation (NLSE) also supports the existence of solitons.

Meanwhile, in 1972, A. Hasegawa and F. D. Tappert showed that the NLSE is the appropriate equation to describe nonlinear optical light pulse propagation through optical fibers. At Bell Labs, L. Mollenaure, R. Stolen, and J. Gordon were able to observe optical soliton propagation in 1980. At this stage, a remarkable suggestion was put forward by Hasegawa and Tappert: the possibility of a soliton-based all-optical communication system. This idea stimulated many researchers, and numerous research papers in this field started appearing in different journals. Meanwhile, optoelectronics and optical communication technology progressed rapidly, largely as a result of the information revolution. Due to rapidly growing demand from the industry, business, commerce, education, entertainment, and government sectors, larger and larger quantities of information were being gathered, transmitted, and processed. One important and essential ingredient in this information revolution was optical fiber communication systems, which became more and more sophisticated. Coherent systems and soliton-based systems will quite soon be contributing to these developments. In order to design such systems, knowledge in mathematical formalism of solitons is essential.

In view of this background, chapter 1 introduced a brief history of optical solitons. An abridged description of optical waveguides and fibers seems to be helpful for physical understanding of the fascinating world of optical solitons. This is done in a lucid way and requires no prerequisite knowledge of the topic. The enormous advantages of optical communication where optical solitons find wide applications was also highlighted.

In this book, fundamental properties of optical solitons in non-Kerr law media have been described. It is now well established that different types of optical nonlinearities can be used to prevent longitudinal and transverse spreading of light pulses leading to optical solitons, which can be used as bits of information in sequential or parallel processing configurations. Therefore, in the beginning of chapter 2, the different types of optical nonlinearities were briefly discussed and their respective analytic forms were introduced. The most important nonlinearity is Kerr law nonlinearity, in which the refractive index is proportional to the first power of light intensity. Obviously, this nonlinearity has drawn wide attention and many researchers have investigated optical solitons in this medium extensively during the last $2\frac{1}{2}$ decades. In the hierarchy of order of attention, parabolic nonlinearity, also known as *cubic-quintic nonlinearity*, comes next. Whereas Kerr law nonlinearity grows with intensity, quintic nonlinearity executes slow saturation in case the quintic term is negative; on the other hand, it shows much faster growth with intensity when this term is positive. However, parabolic nonlinearity with a negative quintic term, also known as *defocusing quintic nonlinearity*, is physically more meaningful. Another important nonlinearity introduced in this chapter is power law nonlinearity, in which the refractive index is proportional to the pth power of light intensity. For a special case of $p = 1$, this type reduces to Kerr nonlinearity. Dual type power law is another form of nonlinearity that, in special cases, reduces to power law or parabolic law nonlinearity. The last form of nonlinearity discussed in this chapter is saturating law nonlinearity. Obviously, this law leads to saturation of the refractive index at large intensities. This law exhibits bistable solitons that can have different peak powers but the same widths. The NLSE and its modified form, which describe optical soliton propagation, possess several integrals of motion. A method to identify a few of them has been identified. The general evolution equations of four soliton parameters—amplitude (or width), frequency, soliton center, and phase—have been introduced. The modified evolution equations due to small perturbations have subsequently been introduced. At this stage, a method to obtain the quasi-stationary (QS) solution has been described. As an example, an application of this method was illustrated.

The concept of optical solitons is based on the integrability of the NLSE. The formation of solitons is a result of balance between group velocity dispersion and optical nonlinearity-induced self-phase modulation. Starting from the first principle, the NLSE was derived for optical fibers in chapter 3. The integrability of the NLSE was demonstrated by Zhakarov and Shabat using the famous inverse scattering transform technique. In a lucid way, this method was introduced and the method for obtaining the one soliton solution of the NLSE was discussed. An important property of the NLSE with Kerr law nonlinearity is that it has an infinite number of degrees of freedom and, consequently, an infinite number of conserved quantities. In fact, this is considered to be one of the definitions of complete integrability. Four important conserved quantities were outlined: energy(E), linear momentum(M), the Hamiltonian (H), and the Hamiltonian for the first integrable hierarchy of the NLSE (H_1). It should be pointed out that even though the ideal NLSE, in which dispersion is balanced by Kerr law nonlinearity, is integrable, the explicit information that can be obtained from the solution is often rather limited. This situation has prompted an effort to complement the exact analytical solution method with approximate methods that sacrifice exactness in order to obtain explicit results and a clearer picture of the properties of the solution. One such method is the direct variational method based on trial functions. Variational formalism has been used successfully and extensively by several authors to address different nonlinear optical problems involving the NLSE and its modified form. The main advantages of the variational method are its simplicity and capacity to provide a clear, qualitative picture and good qualitative results. This powerful method was introduced in chapter 3. In a real-world situation, in an optical communication system, for example, soliton propagation is affected by different types of perturbation. As a matter of fact, these perturbations could result in loss of soliton amplitude due to attenuation, higher-order dispersion, self-steepening, and other factors. Due to the presence of these perturbations, the parameters that characterize a stable soliton do not remain constant. These soliton parameters are soliton amplitude as well as pulse width, speed as well as frequency, soliton position, and the soliton phase. Therefore, a model to deal with these perturbations is essential. The modified variational formalism or perturbed variational method is in many cases a very powerful method to investigate modified NLSEs (mNLSEs). Thus, a model to deal with these mNLSEs has been introduced at the end of the discussion on variational formalism. This method enables one to derive the evolution equations of the soliton parameters outlined earlier. These evolution equations can be further investigated to reveal soliton properties. In addition, QS solitons have been identified using the multiple-scales perturbation method. This technique was first used in the NLSE by Kodama and Ablowitz in 1981 [228] and was further studied by Biswas, who extended it to cases of non-Kerr law nonlinearities [81, 82, 86, 88]. When the nature of a perturbation is Hamiltonian, an additional powerful method to deal with such problems is the Lie transform. This technique was first applied to integrate the NLSE with Hamiltonian-type perturbations in 1994 by Kodama and colleagues [241]. This method was briefly introduced at the end of the chapter and a comparative study of the multiple-scale techniques and Lie transform technique were succintly discussed.

The dynamics of soliton propagation in power law, parabolic law, and dual-power law were respectively addressed in chapters 4, 5, and 6.

Mathematical formalism of these chapters is identical, though the results are different. In each chapter, the propagation dynamics without perturbation terms was considered first. As usual, several conserved quantities were evaluated. In order to offset loss due to attenuation, linear amplification was introduced. However, since linear amplification leads to unstable growth of soliton amplitude, saturable amplifiers were introduced to compensate for the loss. A model with saturation terms included in the NLSE is more satisfactory from a physical point of view, since stable soliton propagation is ensured in principle over an infinite propagation distance, including transoceanic distances. The QS soliton theory of soliton dynamics was invoked to locate fixed point.

An important class of optical nonlinearity that has drawn much attention recently is saturable law nonlinearity. Two different forms of such nonlinearity are known and studied widely. Both forms reduce to Kerr nonlinearity at low light intensity, whereas at large light intensity, both saturate to a constant value. In chapter 7, the mNLSE was derived to describe stationary wave propagation in such a medium and conserved quantities were evaluated. The existence of bistable solitons, which is an exciting feature of such mNLSEs, was discussed. The mNLSE in saturating media is not integrable, hence variational formalism was invoked. An example of an arbitrary pulse was considered and its dynamics were investigated. A second order nonlinear ordinary differential equation for soliton width was derived. At this stage, a potential formalism to show equivalency of pulse dynamics with that of a nonlinear oscillator was introduced. The potential has been examined in detail to extract important features of solitary waves. Conditions for stationary state were identified, and regions of stable and unstable propagation located. The theoretical treatment was extended to both lossless and lossy media.

In a real communication system, soliton pulses never travel as lone entities. In fact, a continuous stream of optical pulses travels along the line. Since a soliton solution ideally allows only one soliton in $(-\infty, \infty)$, an adjacent soliton always modifies the ideal soliton solution. This modification induces an interaction force between neighboring solitons. Therefore, interaction among pulses of the same channel, as well as among pulses of different channels, is extremely important. In fact, strength of this interaction limits transmission capacity. In order to avoid soliton interaction, two adjacent solitons should be separated. This requirement is a drawback for soliton communication systems. In chapter 8, following the footsteps of Karpman and Solovev formalism that was first introduced in 1981 [213], the quasi-particle theory was developed to study the soliton–soliton interaction of two neighboring pulses in the same channel. Both Hamiltonian and non-Hamiltonian perturbations were considered. For example, the perturbation terms due to linear damping or gain, saturable amplification, third and fourth order dispersion, band pass filter, nonlinear dispersion, and self-steeping were incorporated. In order to avoid repetition of identical mathematical development, instead of considering three types of nonlinearity in three different chapters, Kerr law, power law, parabolic law, and dual-power law nonlinearities are treated in this chapter in a compact form. After developing appropriate dynamical equations for four important pulse parameters, direct numerical simulation of the NLSE was carried out. Numerical simulation results have identified certain perturbation terms that are helpful for enhancement of collision distance. For example, the introduction of a filter in the NLSE plays this role quite efficiently.

The focus of chapter 9 was the effect of stochastic perturbation of soliton propagation. In many cases, random noise affects the performance of soliton propagation. This random noise may arise due to amplified spontaneous emission from an amplifier. Moreover, random nonuniformities in optical fibers, such as fluctuations in the values of the dielectric constant, may also act as sources of random noise. Keeping these factors in mind, the perturbed NLSE was introduced to incorporate the random effects discussed in this chapter. The perturbed NLSE was studied using soliton perturbation theory to derive the corresponding Langevin equations for four types of nonlinearity: Kerr, power, parabolic- and dual-power law. These Langevin equations were further pursued to extract system information.

Optical couplers and switches are essential components of optical communication systems. In chapter 10, such couplers and switches were described. The coupled NLSE for twin-core fibers was introduced at the beginning of the chapter. A couple of conserved quantities were identified, parameter dynamics were worked out. Next, multiple cores were introduced. Here, two different cases were considered. In one, coupling with nearest neighbors was discussed, while in the second case, coupling with all neighbors was discussed. At the end of the chapter, a mathematical theory of the coupled magneto-optic waveguides was introduced.

The last chapter, chapter 11, talks very briefly about the existence of optical bullets, which are optical solitons in $1 + 3$ dimensions. The NLSE in $1 + 3$ dimensions was introduced, its conserved quantities were studied, and the basic parameter dynamics were obtained.

Hints and Solutions

Chapter 2

1. Perform the operation $q_x^* \times \frac{\partial}{\partial x}(2.1) - q_x \times \frac{\partial}{\partial x}$ (2.23).
2. Differentiate q given by (2.36) with respect to t and subtract it from its conjugate. Make use of the NLSE given by (2.1).

Chapter 3

1. Use the soliton perturbation theory formulae given by (3.97) and (3.98). For velocity, use the formula given by (3.104).

Chapter 4

1. Use the formulae for soliton perturbation theory given by (4.21) and (4.23).
2. Use the formula for the velocity of the soliton given by (4.29).
3. The fixed point is obtained by setting the right-hand sides of the dynamical system to zero.

Chapter 5

1. Use l'Hospitals' rule to evaluate the limits.
2. For $v = 0$, $a = 1$. Use the fact that $1 + \cosh\phi = 2\cosh^2\phi/2$. Also, make use of the fact that $B = A\sqrt{2}$. Finally, define $A/\sqrt{2}$ as A of the Kerr law soliton.

3. Use the formulae for soliton perturbation theory that are given by (5.21) and (5.23).

4. Use the formula for the velocity that is given by (5.30).

5. Use Rabbe's test for the convergence of a series from Real Analysis.

Chapter 6

1. If $v \longrightarrow 0$, then $a \longrightarrow 1$ in (6.11). Now, use the double angle formula for the cosh function, as in exercise 5.2.

2. Expand Gauss' hypergeometric function and take the approximation as $v \longrightarrow 0$.

3. Consider both right-hand as well as left-hand limits.

4. Use the soliton perturbation theory formulae given by (6.22) and (6.24).

5. Use the formula for the velocity given by (6.31).

Chapter 7

1. Use the definition of the Frechet derivative that is given in (2.35).

Chapter 9

1. Use the adiabatic parameter dynamics formulae given by (9.11) and (9.13), respectively. Note that for Kerr law nonlinearity, $E = 2A$.

2.

$$\langle \kappa(t)\kappa(t') \rangle = -\frac{D}{1-2D}\left\{e^{[-(1-D)|t-t'|]} - e^{[-(1-2D)(t+t')-D|t-t'|]}\right\} + O(D^2)$$

Bibliography

1. F. Abdullaev, S. Darmanyan & P. Khabibullaev. *Optical Solitons*. Springer Verlag, New York, NY. (1993).
2. F. Abdullaev. *Theory of Solitons in Inhomogenous Media*. John Wiley and Sons, New York, NY. (1994).
3. F. K. Abdullaev & J. Garnier. "Solitons in media with random dispersive perturbations," *Physica D*. Vol 134, Issue 3, 303–315. (1999).
4. F. K. Abdullaev, J. C. Bronski & G. Papanicolaou. "Soliton perturbations and the random Kepler problem," *Physica D*. Vol 135, Issue 3-4, 369–386. (2000).
5. M. J. Ablowitz & H. Segur. *Solitons and the Inverse Scattering Transform*. SIAM, Philadelphia, PA. (1981).
6. M. J. Ablowitz & P. A. Clarkson. *Solitons, Nonlinear Evolution Equations and Inverse Scattering*. Cambridge University Press, Cambridge. (1993).
7. M. J. Ablowitz, B. Prinari & D. Trubatch. *Discrete and Continuous Nonlinear Schrödinger Systems*. Cambridge University Press, Cambridge. (2004).
8. M. J. Ablowitz, G. Biondini & L. A. Ostrovsky. "Optical solitons: Perspectives and applications," *Chaos*. Vol 10, Issue 3, 471–474. (2000).
9. M. J. Ablowitz, G. Biondini & S. Blair. "Localized multi-dimensional optical pulses in non-resonant quadratic materials," *Mathematics and Computers in Simulation*. Vol 56, Issue 6, 511–519. (2001).
10. A. B. Aceves, C. De Angelis, A. M. Rubenchik & S. K. Turitsyn. "Multidimensional soliton in fiber arrays," *Optics Letters*. Vol 19, Issue 5, 329–331. (1994).
11. A. B. Aceves, C. D. Angelis, G. Nalesso & M. Santagiustina. "Higher-order effects in bandwidth-limited soliton propagation in optical fibers," *Optics Letters*. Vol 19, 2104–2106. (1994).
12. A. B. Aceves, G. G. Luther, C. D. Angelis, A. M. Rubenchik & S. K. Turitsyn. "Optical pulse compression using fiber arrays," *Optical Fiber Technology*. Vol 1, Issue 3, 244–246. (1995).
13. A. B. Aceves, C. D. Angelis, G. G. Luther, A. M. Rubenchik & S. K. Turitsyn. "All-optical-switching and pulse amplification and steering in nonlinear fiber arrays," *Physica D*. Vol 87, 262–272. (1995).
14. A. B. Aceves, G. G. Luther, C. D. Angelis, A. M. Rubenchik & S. K. Turitsyn. "Energy localization in nonlinear fiber arrays: Collapse-effect compressor," *Physical Review Letters*. Vol 75, Issue 1, 73–76. (1995).
15. A. B. Aceves, C. D. Angelis, T. Peschel, R. Muschall, F. Lederer, S. Trillo & S. Wabnitz. "Discrete self-trapping, soliton interactions and beam steering in nonlinear waveguide arrays," *Physical Review E*. Vol 53, 1172–1189. (1996).
16. A. B. Aceves. "Optical gap solitons: Past, present and future; theory and experiments," *Chaos*. Vol 10, Issue 3, 584–589. (2000).
17. V. V. Afanasjev. "Interpretation of the effect of reduction of soliton interaction by bandwidth-limited amplification," *Optics Letters*. Vol 18, 790–792. (1993).
18. V. V. Afanasjev, J. S. Aitchison & Y. S. Kivshar. "Splitting of high-order spatial solitons under the action of two-photon absorption," *Optics Communications*. Vol 116, 331–338. (1995).

19. V. V. Afanasjev, N. N. Akhmediev & J. M. Soto-Crespo. "Three forms of localized solutions of the quintic complex Ginzburg-Landau equation," *Physical Review E*. Vol 53, Issue 2, 1931–1939. (1996).
20. V. V. Afanasjev & N. N. Akhmediev. "Soliton interaction in nonequilibrium dynamical systems," *Physical Review E*. Vol 53, 6471–6475. (1996).
21. V. V. Afanasjev, P. L. Chu & Y. S. Kivshar. "Breathing spatial solitons in non-Kerr media," *Optics Letters*. Vol 22, Issue 18, 1388–1390. (1997).
22. G. P. Agrawal. *Nonlinear Fiber Optics*. Elsevier Science, North Holland. (2001).
23. G. P. Agrawal. *Applications of Nonlinear Fiber Optics*. Elsevier Science, North Holland. (2001).
24. G. P. Agrawal & R. W. Boyd. *Contemporary Nonlinear Optics*. Academic Press, San Diego. (1992).
25. N. N. Akhmediev, V. I. Korneev & R. F. Nabiev. "Modulation instability of the ground state of the nonlinear wave equation: Optical machine gun," *Optics Letters*. Vol 15, 393–395. (1992).
26. N. N. Akhmediev, V. V. Afanasjev & J. M. Soto-Crespo. "Singularities and special soliton solutions of the cubic-quintic complex Ginzburg-Landau equation," *Physical Review E*. Vol 53, 1190–1201. (1993).
27. N. N. Akhmediev & J. M. Soto-Crespo. "Generation of a train of three-dimensional nonlinear Schrodinger equation," *Physical Review A*. Vol 47, Issue 2, 1358–1364. (1993).
28. N. N. Akhmediev & A. Ankiewicz. "Novel soliton states and bifurcation phenomena in nonlinear fiber optics," *Physical Review Letters*. Vol 70, Issue 16, 2395–2398. (1993).
29. N. N. Akhmediev & V. V. Afanasjev. "Novel arbitrary amplitude soliton solutions of the cubic-quintic complex Ginzburg-Landau equation," *Physical Review Letters*. Vol 75, Issue 12, 2320–2323. (1995).
30. N. N. Akhmediev, V. V. Afanasjev & J. M. Soto-Crespo. "Singularities and special soliton solutions of the cubic-quintic complex Ginzburg-Landau equation," *Physical Review E*. Vol 53, Issue 1, 1190–1201. (1996).
31. N. N. Akhmediev & A. Ankiewicz. *Solitons: Nonlinear Pulses and Beams*. Chapman and Hall, UK. (1997).
32. N. N. Akhmediev, A. Ankiewicz & J. M. Soto-Crespo. "Multisoliton solutions of the complex Ginzburg-Landau equation," *Physical Review Letters*. Vol 79, 4047–4050. (1997).
33. N. N. Akhmediev. "Spatial solitons in Kerr and Kerr-like media," *Optical and Quantum Electronics*. Vol 30, 535–569. (1998).
34. N. N. Akhmediev, A. Ankiewicz & J. M. Soto-Crespo. "Stable soliton pairs in optical transmission lines and fiber lasers," *Journal of Optical Society of America B*. Vol 15, 515–523. (1998).
35. N. N. Akhmediev, A. Ankiewicz & R. Grimshaw. "Hamiltonian-versus-energy diagrams in soliton theory," *Physical Review E*. Vol 59, Issue 5, 6088–6096. (1999).
36. N. N. Akhmediev & A. Ankiewicz. "Multi-soliton complexes," *Chaos*. Vol 10, Issue 3, 600–612. (2000).
37. N. N. Akhmediev, A. Rodrigues & G. Townes. "Interaction of dual-frequency pulses in passively mode-locked lasers," *Optics Communications*. Vol 187, 419–426. (2001).
38. D. Anderson. "Variational approach to nonlinear pulse propagation in optical fibers," *Physical Review A*. Vol 27, Issue 6, 3135–3145. (1983).

39. D. Anderson & M. Lisak. "Bandwidth limits due to mutual pulse interaction in optical soliton communication systems," *Optics Letters*, Vol 11, Issue 3, 174–176. (1986).

40. D. Anderson, M. Lisak, B. Malomed & M. Quiroga-Teixeiro. "Tunneling of an optical soliton through a fiber junction," *Journal of Optical Society of America B*. Vol 11, 2380–2384. (1994).

41. D. Anderson, M. Lisak & B. A. Malomed. "Three-wave interaction solitons in a dispersive medium with quadratic nonlinearity," *Optics Communications*. Vol 126, 251–254. (1996).

42. D. Anderson, M. Lisak & A. Berntson. "A variational approach to nonlinear evolution equations in optics," *Pramana*. Vol 57, Issue 5 & 6, 917–936. (2001).

43. R. L. Anderson & N. H. Ibragimov. *Lie-Bäcklund Transformations in Applications*. SIAM, Philadelphia, PA. (1979).

44. A. Ankiewicz, N. N. Akhmediev & G. D. Peng. "Stationary soliton states in couplers with saturable nonlinearity," *Optical and Quantum Electronics*. Vol 27, 193–200. (1995).

45. A. Ankiewicz & N. N. Akhmediev. "Analysis of bifurcations for parabolic nonlinearity optical couplers," *Optics Communications*. Vol 124, 95–102. (1996).

46. D. Artigas, L. Torner & N. N. Akhmediev. "Asymmetrical splitting of higher-order solitons induced by quintic nonlinearity," *Optics Communications*. Vol 143, Issue 4–6, 322–328. (1997).

47. M. Arumugam. "Optical fiber communication - An overview." *Pramana*. Vol 57, Issue 5 & 6, 849–869. (2001).

48. A. Arraf, H. He & C. M. de Sterke "Deep gratings with a $\chi^{(2)}$ nonlinearity: Analysis and solutions," *Optical Fiber Technology*. Vol 5, 223–234. (1999).

49. I. B. Bakholdin. "Solitary waves and the structures of discontinuities in non-dissipative models with complex dispersion," *Journal of Applied Mathematics and Mechanics*. Vol 67, Issue 1, 43–56. (2003).

50. G. Baldwin. *An Introduction to Nonlinear Optics*. Kluwer Academic Publishers, Boston. (1969).

51. O. Bang, J. J. Rasmussen & P. L. Christiansen. "Subcritical localization in the discrete nonlinear Schrödinger equation with arbitrary power nonlinearity," *Nonlinearity*. Vol 7, Issue 1, 205–218. (1994).

52. O. Bang, P. L. Christiansen, F. If, K. O. Rasmussen & Y. B. Gaididei. "White noise in the two-dimensional nonlinear Schrödinger equation," *Applicable Analysis*. Vol 57, 3–15. (1995).

53. O. Bang, Y. S. Kivshar & A. V. Buryak. "Bright spatial solitons in defocusing Kerr media supported by cascaded nonlinearities," *Optics Letters*. Vol 22, Issue 22, 1680–1682. (1997).

54. O. Bang, L. Berge & J. J. Rasmussen. "Fusion, collapse and stationary bound states of incoherently coupled waves in bulk cubic media," *Physical Review E*. Vol 59, Issue 4, 4600–4613. (1999).

55. O. Bang, W. Krolikowski, J. Wyller & J. J. Rasmussen. "Collapse arrest and soliton stabilization in nonlocal nonlinear media," *Physical Review E*. Vol 66, 046619, 4 pages. (2002).

56. B. Basu-Mallick & T. Bhattacharyya. "Algebraic Bethe ansatz for a quantum integrable derivative nonlinear Schrödinger model," *Nuclear Physics B*. Vol 634, Issue 3, 611–627. (2002).

57. B. Basu-Mallick & T. Bhattacharyya. "Jost solutions and quantum conserved quantities of an integrable derivative nonlinear Schrödinger model," *Nuclear Physics B*. Vol 668, Issue 3, 415–446. (2003).

58. T. B. Benjamin. "The stability of solitary waves," *Proceedings of Royal Society of London A*. Vol 328, 153–183. (1972).

59. L. Berge, O. Bang, J. J. Rasmussen & V. K. Mezentsev. "Self-focusing and solitonlike structures in materials with competing quadratic and cubic nonlinearities," *Physical Review E*. Vol 55, Issue 3, 3555–3570. (1997).

60. X. Bingzhen & W. Wenzheng. "Traveling-wave method for solving the modified nonlinear Schrödinger equation describing soliton propagation along optical fibers," *Physical Review E*. Vol 51, Issue 2, 1493–1498. (1995).

61. A. Biswas. "Optical soliton perturbation with bandwidth limited amplification and saturable amplifiers," *Journal of Nonlinear Optical Physics and Applications*. Vol 8, Issue 2, 277–288. (1999).

62. A. Biswas. "Soliton-soliton interaction in optical fibers," *Journal of Nonlinear Optical Physics and Materials*. Vol 8, Issue 4, 483–495. (1999).

63. A. Biswas. "Integro-differential perturbation of optical solitons," *Journal of Optics A*. Vol 2, Issue 5, 380–388. (2000).

64. A. Biswas. "Perturbation of optical solitons and quasi-solitons," *Journal of Electromagnetic Waves and Applications*. Vol 14, Issue 1, 95–114. (2000).

65. A. Biswas. "Solitons in nonlinear fiber arrays," *Journal of Electromagnetic Waves and Applications*. Vol 15, Issue 9, 1189–1196. (2001).

66. A. Biswas. "Dispersion-managed solitons in multiple-core nonlinear fiber arrays," *Fiber and Integrated Optics*. Vol 20, Issue 6, 571–579. (2001).

67. A. Biswas. "Solitons in multiple-core couplers," *Journal of Nonlinear Optical Physics and Materials*. Vol 10, Issue 3, 329–336. (2001).

68. A. Biswas. "Optical soliton perturbation with nonlinear damping and saturable amplifiers," *Mathematics and Computers in Simulation*. Vol 56, Issue 3, 521–537. (2001).

69. A. Biswas. "Optical soliton perturbation with higher-order dispersions," *Fiber and Integrated Optics*. Vol 20, Issue 2, 171–189. (2001).

70. A. Biswas. "Optical soliton perturbation with Raman scattering and saturable amplifiers," *Optical and Quantum Electronics*. Vol 33, 289–304. (2001).

71. A. Biswas. "Dynamically stable solitons in optical fibers," *Fiber and Integrated Optics*. Vol 20, Issue 6, 617–624. (2001).

72. A. Biswas. "Perturbation of solitons due to power law nonlinearity," *Chaos, Solitons and Fractals*. Vol 12, Issue 3, 579–588. (2001).

73. A. Biswas & A. B. Aceves. "Dynamics of solitons in optical fibers," *Journal of Modern Optics*. Vol 48, Issue 7, 1135–1150. (2001).

74. A. Biswas. "Perturbation of solitons with non-Kerr law nonlinearity," *Chaos, Solitons and Fractals*. Vol 13, Issue 4, 815–823. (2002).

75. A. Biswas. "Integro-differential perturbation of non-Kerr law solitons," *Chaos, Solitons and Fractals*. Vol 14, Issue 4, 673–679. (2002).

76. A. Biswas. "Optical soliton perturbation with Raman scattering and nonlinear damping," *Fiber and Integrated Optics*. Vol 21, 125–143. (2002).

77. A. Biswas. "Multiple-scale analysis for non-Kerr law solitons," *International Mathematical Journal*. Vol 2, Issue 12, 1157–1197. (2002).

78. A. Biswas. "Theory of optical bullets," *Progress in Electromagnetic Research*. Vol 36, 21–59. (2002).

79. A. Biswas. "Theory of optical couplers," *Optical and Quantum Electronics*. Vol 35, Issue 3, 221–235. (2003).
80. A. Biswas. "Dispersion-managed solitons in optical couplers," *Journal of Nonlinear Optical Physics and Applications*. Vol 12, Issue 1, 45–74. (2003).
81. A. Biswas. "Quasi-stationary optical solitons with power law nonlinearity," *Journal of Physics A*. Vol 36, Issue 16, 4581–4589. (2003).
82. A. Biswas. "Quasi-stationary optical solitons with parabolic law nonlinearity," *Optics Communications*. Vol 216, Issue 4–6, 427–437. (2003).
83. A. Biswas. "Optical solitons: Quasi-stationarity versus Lie transform," *Optical and Quantum Electronics*. Vol 35, Issue 10, 979–998. (2003).
84. A. Biswas. "Solitons in magneto-optic waveguides," *Applied Mathematics and Computation*. Vol 153, Issue 2, 387–393. (2004).
85. A. Biswas. "Theory of non-Kerr law solitons," *Applied Mathematics and Computation*. Vol 153, Issue 2, 369–385. (2004).
86. A. Biswas. "Quasi-stationary non-Kerr law optical solitons," *Optical Fiber Technology*. Vol 9, Issue 4, 224–259. (2003).
87. A. Biswas. "Adiabatic dynamics of non-Kerr law solitons," *Applied Mathematics and Computation*. Vol 151, Issue 1, 41–52. (2004).
88. A. Biswas. "Quasi-stationary optical solitons with dual-power law nonlinearity," *Optics Communications*. Vol 235, Issue 1–3, 183–194. (2004).
89. A. Biswas. "Stochastic perturbation of optical solitons in Schrödinger-Hirota equation," *Optics Communications*. Vol 239, Issue 4–6, 461–466. (2004).
90. A. Biswas. "Optical soliton perturbation with non-Kerr law nonlinearities," *Progress in Electromagnetic Research*. Vol 50, 231–266. (2005).
91. A. B. Blagoeva, S. G. Dinev, A. A. Dreischuh & A. Naidenov. "Light bullets formation in a bulk media," *IEEE Journal of Quantum Electronics*. Vol QE-27, 2060–2062. (1991).
92. N. Bloembergen. *Nonlinear Optics*. World Scientific Publishing Company, River Edge, NJ. (1996).
93. K. J. Blow & N. J. Doran. "Bandwidth limits of nonlinear (soliton) optical communication systems," *Electronics Letters*. Vol 19, 429–430. (1983).
94. K. J. Blow & N. J. Doran. "Nonlinear limits on bandwidth at the minimum dispersion in optical fibers," *Optics Communications*. Vol 48, 181–184. (1983).
95. K. J. Blow, N. J. Doran & D. Wood. "Suppression of the soliton self-frequency shift by bandwidth-limited amplification," *Journal of Opt Soc Am B*. Vol 5, Issue 6, 1301–1304. (1988).
96. A. D. Boardman, M. Bertolotti & T. Twardowski. *Nonlinear Waves in Solid State Physics*. Kluwer Academic Publishers, Boston. (1991).
97. A. D. Boardman, L. Pavlov & S. Tanev. *Advanced Photonics with Second-Order Optically Nonlinear Processes*. Kluwer Academic Publishers, Boston. (1998).
98. A. D. Boardman & A. P. Suukhorukov. *Soliton-Driven Photonics*. Kluwer Academic Publishers, Boston. (2001).
99. A. D. Boardman & K. Xie "Magneto-optic spatial solitons," *Journal of Optical Society of America B*. Vol 14, Issue 11, 3102–3109. (1995).
100. A. D. Boardman & K. Xie "Spatial bright-dark soliton steering through waveguide coupling," *Optical and Quantum Electronics*. Vol 30, Issue 7, 783–794. (1998).
101. C. M. Bowden & C. D. Cantrell. *Nonlinear Optics and Materials*. SPIE International Society for Optical Engineering, Bellingham, WA. (1991).
102. J. P. Boyd. *Weakly Nonlocal Solitary Waves and Beyond-All-Orders Asymptotics*. Kluwer Academic Publishers, Boston. (1998).

103. R. W. Boyd. *Nonlinear Optics*. Academic Press, San Diego, CA. (2003).
104. H. E. Brandt. *Selected Papers in Nonlinear Optics*. SPIE International Society for Optical Engineering, Bellingham, WA. (1991).
105. J. C. Bronski. "Nonlinear scattering and analyticity properties of solitons," *Journal of Nonlinear Science*. Vol 8, 161–182. (1998).
106. N. Burq, P. Gerard & N. Tzvedkov. "The Cauchy problem for the nonlinear Schrödinger equation on a compact manifold," *Journal of Nonlinear Mathematical Physics*. Vol 10, Supplement 1, 12–27. (2003).
107. S. Burtsev, D. J. Kaup & B. A. Malomed. "Interactions of solitons with a strong inhomogeneity in a nonlinear optical fiber." *Physical Review E*. Vol 52, Issue 4, 4474–4481. (1995).
108. V. S. Busalev & V. E. Grikurov. "Simulation of instability of bright solitons for NLS with saturating nonlinearity," *Mathematics and Computers in Simulation*. Vol 56, Issue 6, 539–546. (2001).
109. V. S. Busalev & C. Sulem. "On asymptotic analysis of solitary waves for nonlinear Schrödinger's equation," *Nonlinear Analysis*. Vol 20, Issue 3, 419–475. (2003).
110. P. N. Butcher & D. Cotter. *The Elements of Nonlinear Optics*. Cambridge University Press, Cambridge. (1991).
111. G. Chaohao. *Soliton Theory and its Applications*. Springer Verlag, New York, NY. (1995).
112. D. Cai, A. R. Bishop, N. Groenbech-Jensen & B. A. Malomed. "Stabilizing a breather in the damped nonlinear Schrödinger equation driven by two frequencies." *Physical Review E*. Vol 49, Issue 2, R1000-R1002. (1994).
113. S. Chávez-Cerda, M. A. Meneses-Nava & J. J. Sánchez-Mondragón. "Breathing pulses and beams in non-linear media," *Optical and Quantum Electronics*. Vol 30, Issue 7–10. (1998).
114. X. J. Chen & J. Yang. "Direct perturbation theory for solitons of the derivative nonlinear Schrödinger equation and the modified nonlinear Schrödinger equation," *Physical Review E*. Vol 65, 066608. (2002).
115. Y. Chen. "Self-trapped light in saturable nonlinear media," *Optics Letters*. Vol 16, Issue 1, 4–6. (1991).
116. P. L. Christiansen, N. Groenbech-Jensen, P. S. Lomdahl & B. A. Malomed. "Oscillations of eccentric pulsons," *Physics Scripta*. Vol 55, 131–134. (1997).
117. P. L. Chu & C. Desem. "Mutual interaction between solitons of unequal amplitudes in optical fibre," *Electronics Letters*. Vol 21, 1133–1134. (1985).
118. P. L. Chu & C. Desem. "Effect of third order dispersion of optical fibre on soliton interaction," *Electronics Letters*. Vol 21, 228–229. (1985).
119. P. L. Chu & B. Wu. "Optical switching in twin-core erbium-doped fibers," *Optics Letters*. Vol 17, 255–257. (1992).
120. P. L. Chu, B. A. Malomed & G. D. Peng. "Soliton switching and propagation in nonlinear fiber couplers: Analytical results," *Journal of Optical Society of America B*. Vol 10, 1379–1385. (1993).
121. P. L. Chu, B. A. Malomed, G. D. Peng & I. Skinner. "Soliton dynamics in periodically modulated directional couplers," *Physical Review E*. Vol 49, Issue 6, 5763–5767. (1994).
122. P. L. Chu, Y. S. Kivshar, B. A. Malomed, G. D. Peng & M. L. Quiroga-Teixeiro. "Soliton controlling, switching and splitting in fused nonlinear couplers," *Journal of Optical Society of America B*. Vol 12, Issue 5, 898–904. (1995).
123. P. L. Chu, B. A. Malomed & G. D. Peng. "Soliton amplification and reshaping in optical fibers with variable dispersion," *Journal of Optical Society of America B*. Vol 13, Issue 8, 1794–1802. (1996).

124. C. B. Clausen, O. Bang & Y. S. Kivshar. "Spatial solitons and induced Kerr effects in quasi-phase-matched quadratic media," *Physical Review Letters.* Vol 78, 4749–4752. (1997)

125. F. Cooper, C. Lucheroni & H. Shepard. "Variational method for studying self-focusing in a class of nonlinear Schrödinger equations," *Physics Letters A.* Vol 170, Issue 3, 184–188. (1992).

126. J. F. Corney & O. Bang. "Solitons in quadratic nonlinear photonic crystals," *Physical Review E.* Vol 64, 047601. (2001).

127. J. F. Corney & O. Bang. "Modulational instability in periodic quadratic nonlinear materials," *Physical Review Letters.* 133901. (2001).

128. J. F. Corney & O. Bang. "Plane waves in periodic, quadratically nonlinear slab waveguides: Stability and exact Fourier structure," *Journal of Optical Society of America B.* Vol 19, Issue 4, 812–821. (2002).

129. L. C. Crasovan, B. A. Malomed & D. Mihalache. "Spinning solitons in cubic-quintic nonlinear media," *Pramana,* Vol 57, Issue 5 & 6, 1041–1059. (2001).

130. A. Das. *Integrable Models.* World Scientific Publishing Company, Teaneck, NJ. (1989).

131. C. De Angelis. "Self-trapped propagation in the nonlinear cubic-quintic Schrödinger equation: A variational approach," *IEEE Journal of Quantum Electronics.* Vol 30, Issue 3, 818–821. (1994).

132. C. Desem & P. L. Chu. "Soliton interactions in the presence of loss and periodic amplification in optical fibers," *Optics Letters.* Vol 12, Issue 5, 349–351. (1987).

133. J. A. DeSanto. *Mathematical and Numerical Aspects of Wave Propagation.* SIAM, Philadelphia, PA. (1998).

134. A. Desyatnikov, A. Maimistov & B. Malomed. "Three-dimensional spinning solitons in dispersive media with cubic-quintic nonlinearity," *Physical Review E.* Vol 61, No 3, 3107–3113. (2000).

135. Q. Ding & Z. Zhu. "On the Gauge equivalent structure of the modified nonlinear Schrödinger equation," *Physics Letters A.* Vol 295, 192–197. (2002).

136. E. V. Doktorov. "The modified nonlinear Schrödinger equation: Facts and artefacts," *European Physical Journal B.* Vol 29, 227–231. (2002).

137. P. G. Drazin & R. S. Johnson. *Solitons: An Introduction.* Cambridge University Press, Cambridge. (1992).

138. P. D. Drummond, K. V. Kheruntsyan & H. He. "Coherent molecular solitons in Bose-Einstein condensates," *Physical Review Letters.* Vol 81, Issue 15, 3055–3058. (1998).

139. P. D. Drummond & H. He. "Optical mesons," *Physical Review A.* Vol 56, Issue 2, 1107–1110. (1997).

140. H. S. Eisenberg, Y. Silberberg, R. Morandotti, A. R. Boyd & J. S. Aitchison. "Discrete spatial optical solitons in waveguide arrays," *Physical Review Letters.* Vol 81, Issue 16, 3383–3386. (1998).

141. H. S. Eisenberg, R. Morandotti, Y. Silberberg, J. M. Arnold, G. Penneli & J. S. Aitchison. "Optical discrete solitons in waveguide arrays. I. Soliton formation," *Journal of Optical Society of America B.* Vol 19, Issue 12, 2938–2944. (2002).

142. J. N. Elgin. "Stochastic perturbations of optical solitons," *Optics Letters.* Vol 18, Issue 1, 10–12. (1993).

143. D. E. Emundson & R. H. Enns. "The particle-like nature of colliding light bullets," *Physical Review A.* Vol 51, Issue 3, 2491–2498. (1995).

144. D. E. Emundson & R. H. Enns. "Bistable light bullets," *Optics Letters.* Vol 17, 586–588. (1992).

145. R. H. Enns & D. E. Emundson. "Guide to fabricating bistable-soliton-supporting media," *Physical Review A*. Vol 47, Issue 5, 4524–4527. (1993).

146. R. H. Enns & S. S. Rangnekar. "Bistable spheroidal optical solitons," *Physical Review A*. Vol 45, Issue 5, 3354–3357. (1992).

147. R. H. Enns & S. S. Rangnekar. "Variational approach to bistable solitary waves in d dimensions," *Physical Review E*. Vol 48, Issue 5, 3998–4007. (1993).

148. R. H. Enns, S. S. Rangnekar & A. E. Kaplan. "Optical switching between bistable soliton states: A theoretical review," *Optical and Quantum Electronics*. Vol 24, 1295–1314. (1992).

149. C. Etrich, U. Peschel, F. Lederer, B. A. Malomed & Y. S. Kivshar. "Origin of the persistent oscillations of solitary waves in nonlinear quadratic media," *Physical Review E*. Vol 54, Issue 4, 4321–4324. (1996).

150. M. Evans & S. Kielich. *Modern Nonlinear Optics, Vol 1*. John Wiley & Sons, Hoboken, NJ. (1997).

151. M. Evans & S. Kielich. *Modern Nonlinear Optics, Vol 2*. John Wiley & Sons, Hoboken, NJ. (1997).

152. M. Evans & S. Kielich. *Modern Nonlinear Optics, Vol 3*. John Wiley & Sons, Hoboken, NJ. (1997).

153. L. D. Faddeev & L. A. Takhtajan *Hamiltonian Methods in the Theory of Solitons*. Springer Verlag, New York. (1987).

154. R. Fedele, H. Schamel, V. I. Karpman & P. K. Shukla. "Envelope solitons of nonlinear Schrödinger equation with an anti-cubic nonlinearity," *Journal of Physics A*. Vol 36, 1169–1173. (2003).

155. M. F. Ferreira, M. V. Facao & S. V. Latas. "Soliton-like pulses in a system with nonlinear gain," *Photonics and Optoelectronics*. Vol 5, 147–153. (1999).

156. M. F. Ferreira, M. V. Facao & S. V. Latas. "Stable soliton propagation in a system with spectral filtering and nonlinear gain," *Fiber and Integrated Optics*. Vol 19, Issue 1, 31–41. (2000).

157. M. F. Ferreira & S. V. Latas. "Timing jitter in soliton transmission with up and down sliding frequency guiding filters," *Journal of Lightwave Technology*. Vol 19, 332–335. (2001).

158. M. F. Ferreira & S. V. Latas. "Soliton stability and compression in a system with nonlinear gain," *Optical Engineering*. Vol 41, 1696–1703. (2002).

159. R. A. Fisher & J. F. Reintjes. *Nonlinear Optics III*. SPIE International Society for Optical Engineering, Bellingham, WA. (1992).

160. A. S. Fokas & V. E. Zakharov. *Important Developments in Soliton Theory*. Springer Verlag, New York. (1993).

161. M. G. Forest, D. W. McLaughlin, D. J. Muraki & O. C. Wright. "Nonfocusing instabilities in coupled, integrable nonlinear Schrödinger pdes," *Journal of Nonlinear Science*. Vol 10, 291–331. (2000).

162. P. L. Francois & T. Georges. "Reduction of averaged soliton interaction forces by amplitude modulation," *Optics Letters*. Vol 18, 583–585. (1993).

163. H. Frauenkron, Y. S. Kivshar & B. A. Malomed. "Multisoliton collisions in nearly integrable systems," *Physical Review E*. Vol 54, R2244–R2247. (1996).

164. L. Gagnon & P. A. Belanger. "Adiabatic amplification of optical solitons," *Physical Review A*. Vol 43, Issue 11, 6187–6193. (1991).

165. S. Gangopadhyay & S. N. Sarkar. "Variational analysis of spatial solitons of power-law nonlinearity," *Fiber and Integrated Optics*. Vol 20, Issue 2, 191–195. (2001).

166. S. Gatz & J. Herrman. "Soliton propagation and soliton collision in double-doped fibers with a non-Kerr-like nonlinear refractive-index change," *Optics Letters*. Vol 17, Issue 7, 484–486. (1992).

167. T. Georges & F. Favre. "Influence of soliton interaction on amplifier noise-induced jitter: A first-order analytical solution," *Optics Letters*. Vol 16, 1656–1658. (1991).

168. T. Georges & F. Favre. "Modulation, filtering and initial phase control of interacting solitons," *Journal of Optical Society of America B*. Vol 10, 1880–1889. (1993).

169. V. S. Gerdjikov, E. V. Doktorov & J. Yang. "Adiabatic interaction of N ultrashort solitons: Universality of the complex Toda chain model," *Physical Review E*. Vol 64, 056617. (2001).

170. J. M. Ghidaglia & J. C. Saut. "Nonexistence of travelling wave solutions to nonelliptic nonlinear Schrödinger equations," *Journal of Nonlinear Science*. Vol 6, Issue 2, 139–145. (1996).

171. C. G. Goedde, W. L. Kath & P. Kumar. "Controlling soliton perturbations with phase-sensitive amplification," *Journal of Optical Society of America B*. Vol 14, Issue 6, 1371–1379. (1997).

172. P. M. Goorjian & Y. Silberberg. "Numerical simulation of light bullets using the full vector time dependent nonlinear Maxwell equations," *Journal of Optical Society of America B*. Vol 14, Issue 11, 3253–3260. (1997).

173. J. P. Gordon. "Interaction forces among solitons in optical fibers," *Optics Letters*. Vol 8, 596–598. (1983).

174. J.P. Gordon & H. A. Haus. "Random walk of coherently amplified solitons in optical fibre transmission," *Optics Letters*. Vol 11, 665–667. (1986).

175. J.P. Gordon & L. F. Mollenaeur. "Effects of fiber nonlinearities and amplifier spacing on ultra long distance transmission," *Journal of Lightwave Technology*. Vol 9, 170–173. (1991).

176. I. S. Gradshteyn & I. M. Ryzhik. *Table of Integrals, Series, and Products*. Academic Press, New York. (2000).

177. V. S. Grigoryan. "Autosoliton in a fiber with distributed saturable amplifiers," *Optics Letters*. Vol 21, Issue 23, 1882–1884. (1996).

178. V. S. Grigor'yan, A. I. Maimistov & Y.M. Skylarov. "Evolution of light pulses in a nonlinear amplifying medium," *Journal of Experimental and Theoretical Physics*. Vol 67, Issue 3, 530–534. (1988).

179. E. M. Gromov. "Short optical solitons in fibers," *Chaos*. Vol 10, Issue 3, 551–558. (2000).

180. Z. Grujic & H. Kalisch. "The derivative nonlinear Schrödinger equation in analytic classes," *Journal of Nonlinear Mathematical Physics*. Vol 10, Supplement 1, 67–71. (2003).

181. Y. Guo, C. K. Kao, E. H. Li & K. S. Chiang. *Nonlinear Photonics: Nonlinearities in Optics, Optoelectronics, and Fiber Communications*. Springer Verlag, New York. (2002).

182. E. Hanamurs, Y. Kawabe & A. Yamanaka. *Quantum Nonlinear Optics*. Springer Verlag, New York. (2002).

183. A. Hasegawa & F. D. Tappert. "Transmission of stationary nonlinear optical pulses in dispersive dielectric fibers. I. Anomalous dispersion," *Applied Physics Letters*. Vol 23, 142–144. (1973).

184. A. Hasegawa & F. D. Tappert. "Transmission of stationary nonlinear optical pulses in dispersive dielectric fibers. I. Normal dispersion," *Applied Physics Letters*. Vol 23, 171–172. (1973).

185. A. Hasegawa & Y. Kodama. *Solitons in Optical Communications*, Oxford University Press, Oxford. (1995).
186. A. Hasegawa. *Physics and Applications of Optical Solitons in Fibers*. Kluwer Academic Publishers, Boston. (1996).
187. A. Hasegawa. *New Trends in Optical Soliton Transmission Systems*. Kluwer Academic Publishers, Boston. (1998).
188. A. Hasegawa. *Massive WDM and TDM Soliton Transmission Sytems*. Kluwer Academic Publishers, Boston. (2000).
189. A. Hasegawa. "A historical review of application of optical solitons for high speed communications," *Chaos*. Vol 10, Issue 3, 475–485. (2000).
190. A. Hasegawa. "Soliton-based ultra-high-speed optical communications," *Pramana*. Vol 57, Issue 5 & 6, 1097–1127. (2001).
191. A. Hasegawa & M. Matsumoto. *Optical Solitons*. Springer Verlag, New York. (2003).
192. A. Hasegawa. "Theory of information transfer in optical fibers: A tutorial review," *Optical Fiber Technology*. Vol 10, Issue 2, 150–170. (2004).
193. H. Hatami-Hansa, P. L. Chu, B. A. Malomed & G. D. Peng. "Soliton compression and splitting in double-core nonlinear optical fibers," *Optics Communications*. Vol 134, 59–65. (1997).
194. K. Hayata & M. Koshiba. "Solution of self-trapped multidimensional optical beams by Galerkin's method," *Optics Letters*. Vol 17, 841–843. (1992).
195. K. Hayata & M. Koshiba. "Bright-dark solitary-wave solutions of a multidimensional nonlinear Schrödinger's equation," *Physical Review E*. Vol 48, Issue 3, 2312–2315. (1993).
196. K. Hayata & M. Koshiba. "Algebraic solitary-wave solutions of a nonlinear Schrodinger's equation," *Physical Review E*. Vol 51, Issue 2, 1499–1502. (1995).
197. G. S. He & S. H. Liu. *Physics of Nonlinear Optics*. World Scientific Publishing Company, River Edge, NJ (1998).
198. H. He, M. Friese, N. R. Heckenberg & H. Rubinsztein-Dunlop. "Direct observation of transfer of angular momentum to absorptive particles from laser beam with a phase singularity," *Physical Review Letters*. Vol 75, Issue 5, 826–829. (1995).
199. H. He, N. R. Heckenberg & H. Rubinsztein-Dunlop. "Optical pulse trapping with higher-order doughnut beams produced using high-efficiency computer generated phase holograms," *Journal of Modern Optics*. Vol 42, Issue 1, 217–223. (1995).
200. H. He, M. J. Werner & P. D. Drummond. "Simultaneous solitary-wave solutions in a nonlinear parametric waveguide," *Physical Review E*. Vol 54, 896–899. (1996).
201. H. He, P. D. Drummond & B. A. Malomed. "Modulational stability in dispersive optical systems with cascaded nonlinearity," *Optics Communications*. Vol 123, 395–402. (1996).
202. H. He & P. D. Drummond. "Ideal soliton environment using parametric bandgaps," *Physical Review Letters*. Vol 78, Issue 23, 4311–4314. (1997).
203. H. He, A. Arraf, C. M. de Sterke, P. D. Drummond & B. A. Malomed. "Theory of modulational instability in Bragg gratings with quadratic nonlinearities," *Physical Review E*. Vol 59, Issue 5, 6064–6068. (1999).
204. J. Herrman. "Propagation of ultrashot light pulses in fibers with saturable nonlinearity in the normal-dispersion region," *Journal of Optical Society of America B*. Vol 8, Issue 7, 1507–1511. (1991).
205. B. Hermansson & D. Yevick. "Numerical investigation of soliton interaction in optical fibers," *Electronics Letters*. Vol 19, 570–571. (1983).

206. E. J. Hinch. *Perturbation Methods*. Cambridge University Press, Cambridge. (1991).

207. T. Iizuka & Y. S. Kivshar. "Optical gap solitons in nonresonant quadratic media," *Physical Review E*. Vol 59, 7148–7151. (1999).

208. E. Infield & G. Rowlands. *Nonlinear Waves, Solitons and Chaos*. Cambridge University Press, Cambridge. (1990).

209. S. Jana & S. Konar "Induced focussing of two laser beams in cubic quintic nonlinear media," *Physica Scripta*. Vol 70, 354–360. (2004).

210. Z. Jovanoski & R. A. Sammut. "Propagation of Gaussian beams in a nonlinear saturable medium," *Physical Review E*. Vol 50, Issue 5, 4087–4093. (1994).

211. Z. Jovanoski & D. R. Rowland. "Variational analysis of solitary waves in a homogeneous cubic-quintic nonlinear medium," *Journal of Modern Optics*. Vol 48, Issue 7, 1179–1193. (2001).

212. V. I. Karpman & E. M. Maslov. "Perturbation theory for solitons," *Journal of Experimental and Theoretical Physics*. Vol 46, 281–291. (1977).

213. V. I. Karpman & V. V. Solov'ev. "A perturbational approach to the two-soliton systems," *Physica D*. Vol 3, Issue 3, 487–502. (1981).

214. V. I. Karpman. "Radiation of solitons described by a high-order cubic nonlinear Schrödinger equation," *Physical Review E*. Vol 62, Issue 4, 5678–5687. (2000).

215. V. I. Karpman & A. G. Shagalov. "Stability of solitons described nonlinear Schrödinger-type equations with higher-order dispersion," *Physica D*. Vol 144, Issue 1–2, 194–210. (2000).

216. D. J. Kaup. "Perturbation theory for solitons in optical fibers," *Physical Review A*. Vol 42, Issue 9, 5689–5694. (1990).

217. D. J. Kaup & A. C. Newell. "An exact soliton for a derivative nonlinear Schrödinger equation," *Journal of Mathematical Physics*. Vol 19, 798–801. (1978).

218. D. J. Kaup, J. El-Reedy & B. A. Malomed. "Effect of chirp on soliton production," *Physical Review E*. Vol 50, Issue 2, 1635–1637. (1994).

219. O. Keller. *Notions and Perspectives of Nonlinear Optics*. World Scientific Publishing Company, River Edge, NJ. (1996).

220. I. C. Khoo. *Nonlinear Optics and Optical Physics*. World Scientific Publishing Company, River Edge, NJ. (1994).

221. Y. S. Kivshar & A. M. Kosevich. "Evolution of a soliton under the action of small perturbations," *JETP Letters*. Vol 37, 648–651. (1983).

222. Y. S. Kivshar & B. A. Malomed. "Many-particle effects in nearly integrable systems," *Physica D*. Vol 24, 125–154. (1987).

223. Y. S. Kivshar & V. V. Konotop. "Solitons in fiber-optic wave-guides with slowly varying parameters," *Kvantovaya Elektron*. Vol 16, 868–871. (1989).

224. Y. S. Kivshar & B. A. Malomed. "Dynamics of solitons in nearly integrable systems" *Reviews of Modern Physics*. Vol 61, Issue 4, 763–915. (1989).

225. Y. S. Kivshar. "Bright and dark spatial solitons in non-Kerr media," *Optical and Quantum Electronics*. Vol 30, 535–569. (1998).

226. Y. S. Kivshar & B. Luther-Davis. "Dark optical solitons: Physics and applications," *Physics Reports*. Vol 298, 81–197. (1998).

227. Y. S. Kivshar & G. P. Agrawal. *Optical Solitons: From Fibers to Photonic Crystals*. Academic Press, Boston. (2003).

228. Y. Kodama & M. J. Ablowitz. "Perturbations of solitons and solitary waves," *Studies in Applied Mathematics*. Vol 64, 225–245. (1981).

229. Y. Kodama & A. Hasegawa. "Amplification and reshaping of optical solitons in glass fiber-III. Amplifiers with random gain," *Optics Letters*. Vol 8, Issue 6, 342–344. (1983).

230. Y. Kodama & K. Nozaki. "Soliton interaction in optical fibers," *Optics Letters*. Vol 12, Issue 12, 1038–1040. (1987).

231. Y. Kodama & A. Hasegawa. "Nonlinear pulse propagation in a monomode dielectric guide," *IEEE Journal of Quantum Electronics*. Vol 23, 510–524. (1987).

232. Y. Kodama & S. Wabnitz. "Reduction of soliton interaction forces by bandwidth limited amplification," *Electronics Letters*. Vol 27, Issue 21, 1931–1933. (1991).

233. Y. Kodama & A. Hasegawa. "Generation of asymptotically stable optical solitons and suppression of the Gordon-Haus effect," *Optics Letters*. Vol 17, 31–33. (1992).

234. Y. Kodama, M. Romagnoli & S. Wabnitz. "Soliton stability and interaction in fiber lasers," *Electronics Letters*. Vol 28, Issue 21, 1981–1983. (1992).

235. Y. Kodama & A. Hasegawa. "Theoretical foundation of optical-soliton concept in fibers," *Progress in Optics*. Vol XXX, No IV, 205–259. (1992).

236. Y. Kodama & A. Hasegawa. "Generation of asymptotically stable optical solitons and suppression of the Gordon-Haus effect." *Optics Letters* Vol 17, 31–33. (1992).

237. Y. Kodama & S. Wabnitz. "Physical interpretation of reduction of soliton interaction forces by bandwidth limited amplification," *Electronics Letters*. Vol 29, Issue 2, 226–227. (1993).

238. Y. Kodama & S. Wabnitz. "Reduction and suppression of soliton interaction by bandpass filters," *Optics Letters*. Vol 18, Issue 16, 1311–1313. (1993).

239. Y. Kodama & S. Wabnitz. "Comment: Physical interpretation of reduction of soliton interaction forces by bandwidth limited amplification," *Electronics Letters*. Vol 29, 226–227. (1993).

240. Y. Kodama & S. Wabnitz. "Analysis of soliton stability and interactions with sliding filters," *Optics Letters*. Vol 19, Issue 3, 162–164. (1994).

241. Y. Kodama, M. Ramagnoli, S. Wabnitz & M. Midrio. "Role of third-order dispersion on soliton instabilities and interactions in optical fibers," *Optics Letters*. Vol 19, 165–167. (1994).

242. S. Konar. "Supergaussian optical beams in higher order nonlinear media," *Nonlinear Optics*. Vol 23, 9–22. (1999).

243. S. Konar & A. Biswas. "Intra-channel collision of Kerr law optical solitons," *Progress in Electromagnetic Research*. Vol 52. (2005).

244. S. Konar & A. Sengupta. "Propagation of an elliptic Gaussian laser beam in a medium with saturable nonlinearity," *Journal of Optical Society of America*. Vol 11, 144–147. (1994).

245. S. Konar & A. Sengupta. "Self-focusing of elliptic Gaussian laser beams in a saturable nonlinear medium," *Indian Journal of Pure and Applied Physics*. Vol 32, 660–666. (1994).

246. S. Konar & A. Sengupta. "Self-focusing of elliptic Gaussian laser beams: Saturable nonlinearity," *Pramana*. Vol 42, 223–228. (1994).

247. S. Konar, P. K. Sen & J. Kumar. "Propagation of elliptic Gaussian laser beams in a cubic quintic nonlinear medium," *Nonlinear Optics*. Vol 19, 291–308. (1999).

248. S. Konar, J. Kumar & P. K. Sen. "Suppression of soliton instability by higher-order nonlinearity in long haul optical communication systems," *Journal of Nonlinear Optical Physics and Materials*. Vol 8, Issue 4, 497–502. (1999).

249. S. Konar. "Decay of periodically amplified solitons and suppression of this decay by fifth order nonlinearity," *Nonlinear Optics*. Vol 24, 277–288. (2000).

250. S. Konar & R. Jain. "Chirped solitons in semiconductor doped glass fibers," *Nonlinear Optics*. Vol 24, 289–309. (2000).

251. S. Konar, P. K. Barhai & S. Medhekar. "Displacement and deflection of optical beams by nonlinear planar waveguide," *Journal of Nonlinear Optical Physics & Materials*. Vol 12, Issue 1, 1–12. (2003).

252. S. Konar & A. Biswas. "Chirped optical pulse propagation in saturating nonlinear media," *Optical and Quantum Electronics*. Vol 36, Issue 10, 905–918. (2004).

253. V. V. Konotop & L. Vazquez. *Nonlinear Random Waves*. World Scientific Publishing Company, River Edge, NJ. (1994).

254. Y. Kosmann-Schwarzbach, B. Grammaticos & K. M. Tamizhmani. *Integrability of Nonlinear Systems*. Springer Verlag, New York. (2004).

255. W. Krolikowski & Y. Kivshar. "Soliton-based optical switching in waveguide arrays," *Journal of Optical Society of America B*. Vol 13, Issue 5, 876–887. (1996).

256. W. Krolikowski, D. Edmundson & O. Bang. "Unified model for partially coherent solitons in logarithmically nonlinear media," *Physical Review E*. Vol 61, Issue 3, 3122–3126. (2000).

257. W. Krolikowski, O. Bang, J. J. Rasmussen & J. Wyller. "Modulational instability in nonlocal nonlinear Kerr media," *Physical Review E*. Vol 64, 016612. (2001).

258. W. Krolikowski & O. Bang. "Solitons in nonlocal nonlinear media: Exact solutions," *Physical Review E*. Vol 63, No 016610. (2001).

259. W. Krolikowski, O. Bang, J. Wyller & J. J. Rasmussen. "Optical beams in nonlocal nonlinear media," *Acta Physica Polonica A*. Vol 103, Issue 2–3, 133–148. (2003).

260. A. Kumar. "Bistable soliton states and switching in doubly inhomogeneously doped fiber couplers," *Pramana*. Vol 57, Issue 5 & 6, 969–979. (2001).

261. A. Kumar, T. Kurz & W. Lauternborn. "Two-state bright solitons in doped fibers with saturating nonlinearity," *Physical Review E*. Vol 53, Issue 1, 1166–1171. (1996).

262. A. Kundu. *Classical and Quantum Nonlinear Integrable Systems: Theory and Application*. Institute of Physics Publishing, Bristol. (2003).

263. T. I. Lakoba & D. J. Kaup. "Perturbation theory for the Manakov soliton and its application to pulse propagation in randomly birefringent fibers," *Physical Review E*. Vol 56, 6147–6165. (1997).

264. M. Lakshmanan & T. Kanna. "Shape changing collision of optical solitons, universal logic gates, and partially coherent solitons in coupled nonlinear Schrödinger equations," *Pramana*. Vol 57, Issue 5 & 6, 885–916. (2001).

265. G. L. Lamb. *Elements of Soliton Theory*. John Wiley & Sons, New York. (1980).

266. B. B. Laud. *Lasers and Nonlinear Optics*. John Wiley & Sons, New York. (1996).

267. J. H. Lee, Y. C. Lee & C. C. Lin. "Exact solutions of DNLS and derivative reaction-diffusion systems," *Journal of Nonlinear Mathematical Physics*. Vol 9, Supplement 1, 87–97. (2002).

268. N. Litchinitser, W. Krolikowski, N. Akhmediev & G. Agrawal. "Asymmetric partially coherent solitons in saturable media," *Physical Review E*. Vol 60, 2377–2380. (1999).

269. C. Lizarraga & A. B. Aceves. "Analysis of switching phenomenon in a dense medium of two level atoms," *Optics Letters*. Vol 18, 687–689. (1993).

270. R. MacKenzie, M. B. Paranjape & W. J. Zakrzewski. *Solitons: Properties, Dynamics, Interactions, Applications*. Springer Verlag, New York. (1999).

271. M. F. Mahmood. "Evolution of optical solitons in nonlinear dispersive lossy fibers," *Optics & Laser Technology*. Vol 33, Issue 6, 379–381. (2001).

272. A. I. Maimistov. "Completely integrable models of nonlinear optics," *Pramana*. Vol 57, Issue 5 & 6, 953–968. (2001).

273. B. A. Malomed. "Waves and solitary pulses in weakly inhomogeneous Ginzburg-Landau equations," *Physical Review E*. Vol 50, 4249–4252. (1994).
274. B. A. Malomed. "Soliton stability in a bimodal optical fiber in the presence of Raman effect," *Physical Review E*. Vol 50, Issue 6, 5142–5144. (1994).
275. B. A. Malomed. "Strong periodic amplification of solitons in a lossy optical fiber: Analytical results," *Journal of Optical Society of America B*. Vol 11, Issue 7, 1261–1266. (1994).
276. B. A. Malomed. "Bound solitons in a nonlinear optical coupler," *Physical Review E*. Vol 51, Issue 2, R864–R866. (1995).
277. B. A. Malomed. "Bound states of envelope solitons," *Physical Review E*. Vol 47, 2874–2880. (1993).
278. B. A. Malomed, G. D. Peng & P. L. Chu. "A nonlinear optical amplifier based on dual core fiber," *Optics Letters*. Vol 21, Issue 5, 330–332. (1996).
279. B. A. Malomed & R. S. Tasgal. "The Raman effect and solitons in an optical fiber with general ellipticity," *Pure and Applied Optics*. Vol 5, 947–965. (1996).
280. B. A. Malomed. "Pulse propagation in nonlinear optical fiber with periodically modulated dispersion: Variational approach," *Optics Communications*. Vol 136, 313–319. (1997).
281. B. A. Malomed, P. Drummond, H. He, D. Anderson, A. Berntson & M. Lisak. "Spatiotemporal solitons in multidimensional optical media with a quadratic nonlinearity," *Physical Review E*. Vol 56, Issue 4, 4725–4735. (1997).
282. B. A. Malomed, M. Goelles, I. M. Uzunov & F. Lederer. "Stability and interaction of pulses in simplified Ginzburg-Landau equation," *Physica Scripta*. Vol 55, 73–79. (1997).
283. B. A. Malomed, G. D. Peng, P. L. Chu, I. Towers, A. V. Buryak & R. A. Sammut. "Stable helical solitons in optical media," *Pramana*. Vol 57, Issue 5 & 6, 1061–1078. (2001).
284. J. T. Manassah, P. L. Baldeck & R. R. Alfano. "Self-focusing, self-phase modulation and diffraction in bulk homogeneous material," *Optics Letters*. Vol 13, 1090–1092. (1988).
285. M. Matsumoto, H. Ikeda, T. Udea & A. Hasegawa. "Stable soliton transmission in the system with nonlinear gain," *Journal of Lightwave Technology*. Vol 13, 658–665. (1995).
286. A. Mecozzi, J. D. Moores, H. A. Haus & Y. Lai. "Soliton transmission control," *Optics Letters*. Vol 16, 1841–1843. (1991).
287. R. Mcleod, K. Wagner & S. Blair. "(3 + 1)-dimensional optical soliton dragging logic." *Physical Review A*. Vol 52, Issue 4, 3254–3278. (1995).
288. S. Medhekar, S. Konar & Rajkamal. "Self tapering and untapering of a self guided laser beam in an absorbing gain medium with nonlinearity," *Pramana*. Vol 44, Issue 3, 249–256. (1995).
289. S. Medhekar, S. Konar & Rajkamal. "Successive untapering and stationary self-trapped propagation of a laser beam in a saturating nonlinear medium," *Laser and Particle Physics*. Vol 13, 559–564. (1995).
290. R. W. Micallaef, V. V. Afanasjev, Y. S. Kivshar & J. D. Love. "Optical solitons with power law asymptotics," *Physical Review E*. Vol 54, 2936–2942. (1996).
291. D. Mihalache, D. Mazilu, L. C. Crasovan, B. A. Malomed & F. Lederer. "Three-dimensional spinning solitons in the cubic-quintic nonlinear medium," *Physical Review E*. Vol 61, Issue 6, 7142–7145. (2000).
292. D. Mihalache, M. Bertolotti & C. Cibilia. "Nonlinear wave propagation in planar structures," *Progress in Optics*. Vol XXVII. 228–309. (1989).

293. D. Mihalache & N. C. Panoiu. "Exact solutions of nonlinear Schrödinger equation for positive group velocity dispersion," *Journal of Mathematical Physics*. Vol 33, Issue 6, 2323–2328. (1992).

294. D. Mihalache & N. C. Panoiu. "Analytic method for solving the nonlinear Schrödinger equation describing pulse propagation in dispersive optic fibers," *Journal of Physics A*. Vol 26, Issue 11, 2679–2697. (1993).

295. D. L. Mills. *Nonlinear Optics: Basic Concepts*. Springer Verlag, New York. (1998).

296. V. P. Mineev. *Topologically Stable Defects and Solitons in Ordered Media*. Taylor and Francis, UK. (1998).

297. L. F. Mollenauer, J. P. Gordon & S. G. Evangelides. "The sliding frequency guiding filter: An improved form of soliton jitter control," *Optics Letters*. Vol 17, 1575–1577. (1992).

298. J. V. Moloney. *Nonlinear Optical Materials*. Springer Verlag, New York. (1998).

299. J. D. Moores. "On the Ginzburg-Landau laser mode-locking model with fifth-order saturable absorber term," *Optics Communications*. Vol 96, Issues 1–3, 65–70. (1993).

300. J. D. Moores, W. S. Wong & H. A. Haus. "Stability and timing maintenance in soliton transmission and storage rings," *Optics Communications*. Vol 113, Issues 1–3, 153–175. (1994).

301. R. Morandotti, U. Peschell, J. S. Aitchison, H. S. Eisenberg & Y. Silberberg. "Dynamics of discrete solitons in optical waveguide arrays," *Physical Review Letters*. Vol 83, Issue 14, 2726–2729. (1999).

302. R. Morandotti, H. S. Eisenberg, Y. Silberberg, M. Sorel & J. S. Aitchison. "Self-focusing and defocusing in waveguide arrays," *Physical Review Letters*. Vol 86, Issue 15, 3296–3299. (2001).

303. J. A. Murdock. *Perturbations: Theory and Methods*. John Wiley & Sons, New York. (1991).

304. R. Muschall, C. Schmidt-Hattenberger & F. Lederer. "Spatially solitary waves in arrays of nonlinear waveguides," *Optics Letters*. Vol 19, Issue 5, 323–325. (1994).

305. M. Nakazawa & H. Kubota. "Physical interpretation of reduction of soliton interaction forces by bandwidth limited amplification," *Electronics Letters*. Vol 28, 958–960. (1992).

306. M. Nakazawa & H. Kubota. "Reply to Kodama and Wabnitz," *Electronics Letters*. Vol 29, 226–227. (1993).

307. M. Nakazawa, H. Kubota, K. Suzuki, E. Yamada & A. Sahara. "Recent progress in soliton transmission technology," *Chaos*. Vol 10, Issue 3, 486–514. (2000).

308. K. Nakkeeran & K. Porsezian. "Solitons in an erbium-doped nonlinear fibre medium with stimulated inelastic scattering," *Journal of Physics A*. Vol 28, Issue 13, 3817–3823. (1995).

309. K. Nakkeeran & K. Porsezian. "Co-existence of a self-induced transparency soliton and a higher order nonlinear Schrödinger soliton in an erbium-doped fiber," *Optics Communications*. Vol 123, Issues 1–3, 169–174. (1996).

310. A. C. Newell. *Solitons in Mathematics and Physics*. SIAM Publishers, Philadelphia. (1985).

311. A. C. Newell & J. V. Moloney. *Nonlinear Optics*. Addison Wesley-Redum City, CA. (1989).

312. P. J. Olver & D. H. Sattinger. *Solitons in Physics, Mathematics, and Nonlinear Optics*. Springer Verlag, New York. (1990).

313. A. Orlowski & K. Sobczyk. "Solitons and shock waves under random external noise," *Reports on Mathematical Physics*. Vol 27, Issue 1, 59–71. (1989).

314. E. A. Ostrovskaya & Y. S. Kivshar. "Multi-hump optical solitons in a saturable medium," *Journal of Optics B*. Vol 1, 77–83. (1999).

315. T. Ozawa. "On the nonlinear Schrödinger equations of derivative type," *Indiana University Math Journal*. Vol 45, 137–163. (1996).

316. S. L. Palacios. "Optical solitons in highly dispersive media with a dual-power nonlinearity law," *Journal of Optics A*. Vol 5, Issue 3, 180–182. (2003).

317. S. L. Palacios. "Two simple ansatze for obtaining exact solutions of high dispersive nonlinear Schrödinger equations," *Chaos Solitons & Fractals*. Vol 19, Issue 1, 203–207. (2004).

318. N-C. Panoiu, I. V. Melnikov, D. Mihalache, C. Etrich & F. Lederer. "Soliton generation from a multi-frequency optical signal," *Journal of Optics B*. Vol 4, R53–R68. (2002).

319. D. E. Pelinovsky, A. V. Buryak & Y. S. Kivshar. "Instability of solitons governed by quadratic nonlinearities," *Physical Review Letters*. Vol 75, 591–595. (1995).

320. D. E. Pelinovsky, V. V. Afanasjev & Y. S. Kivshar. "Nonlinear theory of oscillating, decaying and collapsing solitons in the generalized nonlinear Schrödinger's equation," *Physical Review E*. Vol 53, Issue 2, 1940–1953. (1996).

321. U. Peschel, R. Morandotti, J. M. Arnold, J. S. Aitchison, H. S. Eisenberg, Y. Silberberg, T. Pertsch & F. Lederer. "Optical discrete solitons in waveguide arrays. 2. Dynamic properties," *Journal of Optical Society of America*. Vol 19, Issue 11, 2637–2644. (2002).

322. K. Porsezian. "Soliton propagation in semiconductor-doped glass fibers with higher-order dispersions," *Pure and Applied Optics*. Vol 5, Issue 4, 345–348. (1996).

323. K. Porsezian. "Soliton models in resonant and nonresonant optical fibers," *Pramana*. Vol 57, Issue 5 & 6, 1003–1009. (2001).

324. K. Porsezian, P. Shanmugha Sundaram & A. Mahalingam. "Complete integrability of N-coupled higher-order nonlinear Schrödinger equations in nonlinear optics," *Journal of Physics A*. Vol 32, Issue 49, 8731–8737. (1999).

325. K. Porsezian & V. C. Kuraikose. *Optical Solitons: Theoretical and Experimental Challenges*. Springer Verlag, New York. (2003).

326. K. Porsezian & K. Nakkeeran. "Solitons in random nonuniform erbium doped nonlinear fiber media," *Physics Letters A*. Vol 206, Issue 3–4, 183–186. (1995).

327. K. Porsezian & D. Vijay Emanuel Muthiah. "Soliton pulse compression in nonuniform birefringent fibers," *Journal of Optics A*. Vol 4, Issue 2, 202–207. (2002).

328. M. J. Potasek. "Novel femtosecond solitons in optical fibers, photonic switching, and computing," *Journal of Physics A* Vol 65, Issue 3, 941–953. (1989).

329. D. Pushkarov & S. Tanev. "Bright and dark solitary wave propagation and bistability in the anomalous dispersion region of optical waveguides with third- and fifth-order nonlinearities," *Optics Communications*. Vol 124, 354–364. (1996).

330. K. I. Pushkarov & D. I. Pushkarov. "Soliton solutions in some nonlinear Schrödinger-like equations," *Reports on Mathematical Physics*. Vol 17, Issue 1, 37–40. (1980).

331. S. Raghavan & G. P. Agrawal. "Switching and self-trapping dynamics of Bose-Einstein solitons," *Journal of Modern Optics*. Vol 47, Issue 7, 1155–1169. (2000).

332. P. M. Ramos & C. R. Pavia. "Self-routing switching of solitonlike pulses in multiple-core nonlinear fiber arrays," *Journal of Optical Society of America B*. Vol 17, Issue 7, 1125–1133. (2000).

333. K. O. Rasmussen, Y. B. Gaididei, O. Bang & P. L. Christiansen. "The influence of noise on critical collapse in the nonlinear Schrödinger equation," *Physics Letters A*. Vol 204, 121–127. (1995).

334. S. G. Rautian. *Nonlinear Optics*. Nova Science. (1992).

335. M. Remoissenet. *Waves Called Solitons: Concepts and Experiments*. Springer Verlag, New York. (1994).

336. P. Roussignol, D. Ricard, J. Lukasik & C. Flytzanis. "New results on optical phase conjugation in semiconductor-doped glasses," *Journal of Optical Society of America B*. Vol 4, Issue 1, 5–13. (1987).

337. K. Rypdal & J. J. Rasmussen. "Blow-up in nonlinear Schrödinger equations II," *Physica Scripta*. Vol 33, 498–504. (1986).

338. K. Rypdal & J. J. Rasmussen. "Stability of solitary structures in nonlinear Schrödinger equations," *Physica Scripta*. Vol 40, 192–201. (1989).

339. K. Rypdal, J. J. Rasmussen & K. Thomsen. "Singularity structure of wave collapse," *Physica D*. Vol 16, 339–359. (1985).

340. M. Schubert & B. Wilhelmi. *Nonlinear Optics and Quantum Electronics*. John Wiley & Sons, New York. (1996).

341. E. G. Sauter. *Nonlinear Optics*. John Wiley & Sons, New York. (1996).

342. M. Segev. "Optical spatial solitons," *Optical and Quantum Electronics*. Vol 30, Issue 7/10, 503–533. (1998).

343. K. Senthilnathan & K. Porsezian. "Evolution of polarization of a nonlinear pulse in birefringent fiber with quintic effects," *Physics Letters A*. Vol 301, Issues 5–6, 433–441. (2002).

344. I. V. Shadrivov, A. A. Sukhorukov & Y. S. Kivshar. "Beam shaping by a periodic structure with negative refraction," *Applied Physics Letters*. Vol 82, 3820–3822. (2003).

345. J. Shatah. "Global existence of small solutions to nonlinear evolution equations," *Journal of Differential Equations*. Vol 46, 409–425. (1982).

346. V. S. Shchesnovich & E. V. Doktorov. "Perturbation theory for the modified nonlinear Schrödinger solitons," *Physica D*. Vol 129, Issue 1–2, 115–129. (1999).

347. Y. R. Shen. *The Principles of Nonlinear Optics*. John Wiley & Sons, New York. (2002).

348. E. Shiojiri & Y. Fuji. "Transmission capability of an optical fibre communication system using index nonlinearity," *Applied Optics*. Vol 24, 358–360. (1985).

349. Y. Silberberg. "Collapse of optical pulses," *Optics Letters*. Vol 15, 1282–1284. (1990).

350. L. Singh, S. Konar & A. K. Sharma. "Resonant cross modulation of two laser beams in a semiconductor slab," *Journal of Physics D*. Vol 34, 2237–2239. (2001).

351. L. Singh, S. N. Rai, S. Konar & A. K. Sharma. "Information exchange between Gaussian laser beams in n-InSb," *Physica Scripta*. Vol 65, 1–4. (2002).

352. V. Singh, B. Prasad & S. P. Ojha. "A comparative study of the modal characteristics and waveguide dispersion of optical waveguides with three different closed loop cross-sectional boundaries," *Optik*. Vol 115, Issue 6, 281–288. (2004).

353. D. V. Skryabin & W. J. Firth. "Dynamics of self-trapped beams with phase dislocation in saturable Kerr and quadratic nonlinear media," *Physical Review E*. Vol 58, 3916–3930. (1998).

354. A. W. Snyder & D. J. Mitchell. "Spatial solitons of the power-law nonlinearity," *Optics Letters*. Vol 18, Issue 2, 101–103. (1993).

355. M. S. Sodha, S. Medhekar, S. Konar, A. Saxena & Rajkamal. "Absorption/amplification induced self tapering and untapering of a laser beam in a saturable nonlinear medium; large nonlinearity," *Optics Letters*. Vol 19, 15–21. (1994).

356. J. M. Soto-Crespo, N. Akhmediev & A. Ankiewicz. "Soliton propagation in optical devices with two-component fields: A comparative study," *Journal of Optical Society of America B*. Vol 12, Issue 6, 1100–1109. (1995).

357. J. M. Soto-Crespo, N. Akhmediev & V. V. Afanasjev. "Algebraic pulse-like solutions of the quintic complex Ginzburg-Landau equation," *Optics Communications*. Vol 118, 587–593. (1995).

358. J. M. Soto-Crespo, V. Afanasjev, N. Akhmediev & G. Town. "Dual-frequency pulses in fiber lasers," *Optics Communications*. Vol 130, 245–248. (1996).

359. G. I. Stegeman & R. H. Stolen. "Waveguides and fibers for nonlinear optics," *Journal of Optical Society of America B*. Vol 6, Issue 4, 652–662. (1989).

360. C. Sulem & P. L. Sulem. *The Nonlinear Schrödinger's Equation*. Springer Verlag, New York. (1999).

361. A. A. Sukhorukov, Y. S. Kivshar, H. S. Eisenberg & Y. Silberberg. "Spatial optical solitons in waveguide arrays," *IEEE Journal of Quantum Electronics*. Vol 39, 31–50. (2003).

362. S. Tanev & D. I. Pushkarov. "Solitary wave propagation and bistability in the normal dispersion region of highly nonlinear optical fibers and waveguides," *Optics Communications*. Vol 141, Issue 5–6, 322–328. (1997).

363. S. Trillo, S. Wabnitz, E. M. Wright & G. I. Stegeman. "Soliton switching in fiber nonlinear directional couplers," *Optics Letters*. Vol 13, 672–674. (1988).

364. R. L. Sutherland. *Handbook of Nonlinear Optics*. Marcel Dekker, New York. (1996).

365. J. M. Soto-Crespo, N. Akhmediev & A. Ankiewicz. "Soliton propagation in optical devices with two-component fields: A comparative study," *Journal of Optical Society of America B*. Vol 12, Issue 6, 1100–1109. (1995).

366. A. W. Snyder & D. J. Mitchell. "Spatial solitons of the power-law nonlinearity," *Optics Letters*, Vol 18, Issue 2, 101–103, (1993).

367. J. M. Soto-Crespo, N. N. Akhmediev & V. V. Afanasjev. "Stability of pulse-like solutions of the quintic complex Ginzburg-Landau equation," *Journal of Optical Society of America B*. Vol 13, 1439–1448. (1996).

368. A. A. Sukhorukov, Y. S. Kivshar, O. Bang & C. M. Soukoulis. "Parametric localized modes in quadratic nonlinear photonic structures," *Physical Review E*. Vol 63, 016615. (2001).

369. V. I. Talanov. "Focusing of light in cubic media," *JETP Letters*. Vol 11, 199–201. (1970).

370. X. Tang & P. Ye. "Calculation of the bit error rate for optical soliton communication systems with lumped amplifiers," *Optics Letters*. Vol 18, 1156–1158. (1993).

371. J. R. Taylor. *Optical Solitons: Theory and Experiment*. Cambridge University Press, Cambridge. (1992).

372. V. Tikhonenko, Y. S. Kivshar, V. V. Steblina & A. A. Zozulya. "Vortex solitons in a saturable optical medium," *Journal of Optical Society of America B*. Vol 15, 79–86. (1998).

373. L. Torner, J. P. Torres, D. V. Petrov & J. M. Soto-Crespo. "From topological charge information to sets of solitons in quadratic nonlinear media," *Optical and Quantum Electronics*. Vol 30, Issue 7/10, 809–827. (1998).

374. L. Torner, D. Mihalache, M. C. Santos & N. N. Akhmediev. "Spatial walking solitons in quadratic nonlinear crystals," *Journal of Optical Society of America B*. Vol 15, Issue 5, 1476–1487. (1998).

375. H. Torress-Silvia & M. Zamorano. "Chiral effects on optical solitons," *Mathematics and Computers in Simulations*. Vol 62, Issue 1–2, 149–161. (2003).

376. S. Trillo & W. Torruellas. *Spatial Solitons*. Springer Verlag, New York. (2001).

377. I. M. Uzunov, V. D. Stoev & T. I. Tsoleva. "N-soliton interaction in trains of unequal soliton pulses in optical fibers," *Optics Letters*. Vol 17, 1417–1419. (1992).

378. I. M. Uzunov, V. D. Stoev & T. I. Tsoleva. "Influence of the initial phase difference between pulses on the N-soliton interaction in trains of unequal solitons in optical fibers," *Optics Communications*. Vol 97, Issue 5–6, 307–311. (1993).

379. I. M. Uzunov, M. Gölles, L. Leine & F. Lederer. "The effect of bandwidth limited amplification on soliton interaction," *Optics Communications*. Vol 110, Issue 3–4, 465–474. (1994).

380. I. M. Uzunov, R. Muschall, M. Gölles, F. Lederer & S. Wabnitz. "Effect of nonlinear gain and filtering on soliton interaction," *Optics Communications*. Vol 118, Issue 5–6, 577–580. (1995).

381. I. Uzunov, R. Muschall, M. Golles, Y. Kivshar, B. Malomed & F. Lederer. "Pulse switching in nonlinear fiber directional couplers," *Physical Review E*. Vol 51, 2527–2537. (1995).

382. I. M. Uzunov, V. Gerdjikov, M. Gölles & F. Lederer. "On the description of N-soliton interactions in optical fibers," *Optics Communications*. Vol 125, Issue 4–6, 237–242. (1996).

383. I. M. Uzunov, M. Golles & F. Lederer. "Enhanced sideband instability in soliton transmissions with semiconductor optical amplifiers," *Optics Letters*. Vol 22, Issue 18, 1406–1408. (1997).

384. S. Wabnitz, Y. Kodama & A. B. Aceves. "Control of optical soliton interactions," *Optical Fiber Technology*. Vol 1, 187–217. (1995).

385. S. Wabnitz. "Control of soliton train transmission, storage, and clock recovery by cw light injection," *Journal of Optical Society of America B*. Vol 13, Issue 12, 2739–2749. (1996).

386. M. Wadati. "Introduction to solitons," *Pramana*. Vol 57, Issue 5 & 6, 841–847. (2001).

387. C. Weilnau, W. Krolikowski, E. A. Ostrovskaya, M. Ahles, M. Geisser, G. McCarthy, C. Denz, Y. S. Kivshar & B. Luther-Davis. "Composite spatial solitons in a saturable nonlinear bulk medium," *Applied Physics B*. Vol 72, 723–727. (2001).

388. M. Weinstein. "Existence and dynamic stability of solitary wave solutions of equations arising in long wave propagations," *Communications in Partial Differential Equations*. Vol 12, Issue 10, 1133–1173. (1987).

389. G. B. Whitham. *Linear and Nonlinear Waves*. John Wiley and Sons, New York. (1999).

390. T. R. Wolinski. "Polarimetric optical fibers and sensors," *Progress in Optics*. Vol XL, 1–75. (2000).

391. J. Wyller & E. Mjolhus. "A perturbation theory for Alfven solitons," *Physica D*. Vol 13, 234–246. (1984).

392. J. Wyller, W. Krolikowski, O. Bang & J. J. Rasmussen. "Generic features of modulational instability in nonlocal Kerr media," *Physical Review E*. Vol 66, 066615. (2002).

393. Z. Xu, L. Li, Z. Li & G. Zhou. "Soliton interaction under the influence of higher-order effects," *Optics Communications*. Vol 210, Issue 3–6, 375–384. (2002).

394. J. Yang & D. J. Kaup. "Stability and evolution of solitary waves in perturbed generalized nonlinear Schrödinger's equation," *SIAM Journal of Applied Mathematics*. Vol 60, Issue 3, 967–989. (2000).

395. V. E. Zakharov & A. B. Shabat. "Exact theory of two-dimensional self-focusing and one-dimensional self-modulation of waves in nonlinear media," *Journal of Experimental and Theoretical Physics*. Vol 34, 62–69. (1972).

396. V. E. Zakharov & V. S. Synakh. "The nature of self-focusing singularity," *Soviet Physics JETP*. Vol 41, 465–468. (1976).

397. V. E. Zakharov & E. A. Kuznetsov. "Optical solitons and quasisolitons," *Journal of Experimental and Theoretical Physics*. Vol 86, Issue 5, 1035–1046. (1998).

398. P. E. Zhidkov. "Existence of solutions to the Cauchy problem and stability of kink-solutions of the nonlinear Schrödinger equation," *Siberian Mathematical Journal*. Vol 33, Issue 2, 239–246. (1992).

399. P. E. Zhidkov. *Korteweg–de Vries and Nonlinear Schrödinger Equations: Qualitative Theory*. Springer Verlag, New York. (2001).

400. C. Zhou, X. T. He & T. Cai. "Pattern structures on generalized nonlinear Schrödinger's equations with various nonlinear terms," *Physical Review E*. Vol 50, Issue 5, 4136–4155. (1994).

Index

A

Advantages of fiber-optic communications, 5–6
Arbitrary pulse propagation, saturable law nonlinearity, 91–98
  lossless uniform media, 94–96
  lossy media, 97–98
  stationary pulse propagation, 96–97

B

Bistable solitons, saturable law nonlinearity, 90–91
Bullets, optical, 161–166
  one plus three dimensions, 162–166
    integrals of motion, 162–164
    parameter evolution, 164–166

C

Couplers, optical, 145–160
  four-port couplers, 146
  functions of, 146–147
  magneto-optic waveguides, 157–160
    mathematical analysis, 158–160
  multiple-core couplers, 152–157
    coupling with all neighbors, 155–157
    coupling with nearest neighbors, 153–155
  optical switching, 147–148
  star coupler, 146
  three-port couplers, 146
  twin-core couplers, 148–152
  types, 146–147
Cubic-quintic nonlinearity. *See* Parabolic nonlinearity

D

Dielectric waveguide, in fiber-optic communications, 6
Dual-power law, 9
  quasi-particle theory, soliton-soliton interaction, 128–134

Hamiltonian perturbations, 131–134
  non-Hamiltonian perturbations, 131
  soliton-soliton interaction, 108–109
  stochastic perturbation, 142–143
Dual-power law nonlinearity, 77–86
  integrals of motion, 79–82
  quasi-stationary solution, 82–85
  traveling wave solution, 78–79

E

Exponential law, 9–10

F

Four-port couplers, 146

G

Graded-index fiber, 6

H

Hamiltonian perturbation, 170
Hamiltonian perturbations, soliton-soliton interaction, 103
Hamiltonian structure, integrals of motion, 41–42
Hasegawa, A., 167
High order polynomial law, 10

I

Integrals of motion
  dual-power law nonlinearity, 79–82
  Kerr law nonlinearity, 39–42
  parabolic law nonlinearity, 70–72
  power law nonlinearity, 59–62
Inverse scattering transform, Kerr law nonlinearity, 33–39
  1-soliton solution, 37–39

195

Printed and bound by CPI Group (UK) Ltd, Croydon, CR0 4YY

23/10/2024

01778239-0006